RED

A *History* of the *Redhead*

JACKY COLLISS HARVEY

BLACK DOG
& LEVENTHAL
PUBLISHERS
NEW YORK

Black Dog & Leventhal Publishers
Hachette Book Group
1290 Avenue of the Americas
New York, NY 10104

www.hachettebookgroup.com
www.blackdogandleventhal.com

Originally published in hardcover by Black Dog & Leventhal Publishers in June 2015
First Trade Paperback edition: March 2018

Black Dog & Leventhal Publishers is an imprint of Hachette Books, a division of Hachette Book Group. The Black Dog & Leventhal Publishers name and logo are trademarks of Hachette Book Group, Inc.

The publisher is not responsible for websites (or their content) that are not owned by the publisher.

The Hachette Speakers Bureau provides a wide range of authors for speaking events. To find out more, go to www.HachetteSpeakersBureau.com or call (866) 376-6591.

Print book interior design by Cindy Joy

Library of Congress Control Number: 2017948542

ISBNs: 978-0-316-47386-6 (trade paperback); 978-1-60376-403-2 (ebook)

Printed in the United States of America

WOR

10 9 8 7 6 5 4 3 2 1

This one is for Mark.

CONTENTS

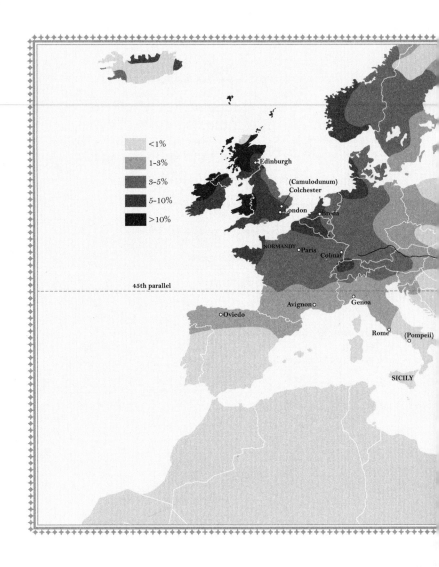

<1%

1-3%

3-5%

5-10%

>10%

Edinburgh

(Camulodunum)
Colchester

London   Breda

NORMANDY   Paris
Colmar

45th parallel

Avignon   Genoa

Oviedo

Rome   (Pompeii)

SICILY

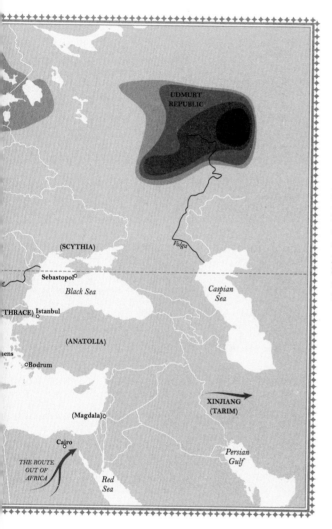

UDMURT
REPUBLIC

(SCYTHIA)

Volga

Sebastopol

Black Sea

Caspian
Sea

(THRACE) Istanbul

(ANATOLIA)

ᴎens

Bodrum

(Magdala)

XINJIANG
(TARIM)

Cairo

Persian
Gulf

THE ROUTE
OUT OF
AFRICA

Red
Sea

**The Redhead
Map of Europe.**

There is a good deal
of controversy over
the accuracy of such
maps, as there is
indeed over so many
issues associated with
red hair, but what it
shows very clearly
is the hotspot in
Russia of the Udmurt
population on the
River Volga and the
increasing frequency
of red hair the farther
north and west
you go, whether in
Scandinavia, Iceland,
the British Isles, or
Ireland.

INTRODUCTION

The study of hair, I found out, does not *take you to*
the superficial edge of our society, the place where
everything silly and insubstantial must dwell.
It takes you, instead, to the centre of things.

GRANT McCRACKEN, *BIG HAIR*, 1995

I am the only redhead in my family, a situation with which many a redhead will be familiar. My mother, now gray-haired, was blonde (was still a blonde, well into her seventies). My father's hair was dark brown. My brother is also blond. My brother's kids have hair that shades from brownish to blondish to positively Aryan. Yet mine is red. When I was little, it was the same orange color as the label on a bottle of Worcestershire sauce; with age it has toned down, to a proper copper. It is not carrots, nor ginger, nor the astonishing fuzz of paprika I remember on the head of a girl at school, a child with skin so white it was almost luminous. I'm not quite at that end of the spectrum, but I am red. It is, with me, as with many other redheads, the single most significant characteristic of my life. If that sounds a little extreme to you, well, you're obviously not a redhead, are you?

Red hair is a recessive gene, and it's rare. Worldwide, it occurs in only 2 percent of the population, although it is slightly more common (2 to 6 percent) in northern and western Europe, or in those with that ancestry (see the map on pages viii-ix).[1] In the great genetic card game, the shuffling of the deck that has made us all, red hair is the two of clubs. It is trumped by every other card in the pack. Therefore, for a red-haired child to result, both parents have to carry the gene, which, blond- or brown-haired as they may very well be, they can be carrying completely unaware. So when a baby appears with that telltale tint to its peach fuzz, expect many jokes and much hilarity. For all my toddlerhood, my mother would blithely ascribe my red hair to either her craving for tomato juice during her pregnancy or to a mysterious redheaded milkman. My grandmother, meanwhile, was fond of quoting the wise old saying that "God gives a woman red hair for the same reason He gives a wasp stripes." But then she was a native of Hampshire, a West Country girl, where redheads were once also known, charmingly, as Dane's bastards, so really, that was letting me off lightly.

I was five before I realized there might be more to being a red-head than incomprehensible teasing by adults. My village school in Suffolk was terrorized by a kindergarten Caligula, a bully from day one, whom we'll call Brian. The rest of us five-year-olds watched in disbelieving horror as Brian roamed about the playground, dispensing armlocks, yanking out hair by the roots, and knocking down birds' nests and laughing as he stamped on the eggs, or fledglings, inside. His genius was to find the thing most precious to you and

---

1 By comparison, somewhere between 16 percent and 17 percent of the population of the planet has blue eyes and 10 to 12 percent of the population is left-handed. Roughly 1 in 10 Caucasian men are born with some degree of color blindness, while 1.1 percent of all births worldwide are of twins. The incidence of albinism, worldwide, is roughly 0.006 percent.

destroy it. One afternoon at the end of a school day he came up behind my friend Karen, who was sporting a new woolly hat of a pale and pretty blue with a large fluffy bobble on the top. Brian seized the hat from Karen's head, ripped off the bobble, and threw it to the ground.

I can still summon up the extraordinary feeling of liberation as the red mists descended. I wound up my right arm like Popeye and punched Brian in the face.

It was a fantastic blow. Brian was knocked flat. As he made to get to his feet, his eye was already swelling shut. Most incredible of all, Brian was in tears. Only then did I realize that my David-and-Goliath moment had been witnessed by all the mothers arriving at the school gate to collect their children, my mother included.

One did not punch. I knew this from the number of times I'd been told off for fighting with my own younger brother. I imagined my punishment. I awaited my mum's reaction, and the reactions of the other mothers at the gate. I was proudly unrepentant, but I knew I was also in any amount of trouble.

The punishment never came. Someone—one of the teachers, I think—picked Brian up and brushed him down. There was laughter. There was an air, astonishingly, of adult approval. My mum, who seemed embarrassed, took my hand and began to hustle me down the road. "Well, what did he expect?" one of her friends remarked, above my head. "She's a redhead!"

She's a redhead. I was five years old, and I had just learned two very important lessons. One, that the world has expectations of redheads, and two, that those expectations give you a license not granted to blondes or brunettes. I was expected to lose my temper.

I dutifully produced appalling tantrums as a child. I was meant to be confident, assertive, and, if I wished, slightly kooky. I could be a screwball. I could be fiery. As I grew older, the list of things I was allowed to do, simply because of the color of my hair, increased. I was allowed to be impulsive. I was allowed to be hot-blooded and passionate (once I reached the age for boyfriends and relationships, it seemed I was almost required so to be). The assumptions and expectations the world made about me and my fellow redheads were endless. I must be Irish. Or Scottish. I must be artistic. I must be spiritual. Was I by any chance psychic? And I must be good in bed. There's a point where all those "musts" start taking on the tone of a command. *She's a redhead.* That was all the world need know, apparently, to know me.

I grew up, and the world got bigger, too. I taught English to a brother and sister from Sicily who were even redder haired and paler skinned and bluer eyed than I am. How did that happen? I traveled farther. I discovered new attitudes toward my red hair, not the same at all as those I had grown up with. Yet the common denominator in every reaction I experienced was this: redheads were viewed as being different. And there has, of course, to be a point when you start asking yourself *why*. Why these assumptions? What's their basis? Do they even have one? Why do they differ from one country to another? Why have they changed, or why have they not, from one century to the next? Where do redheads come from, anyway?

The term "redde-headed" as a synonym for red hair can be tracked back at least to 1565, when it appears in *Thomas Cooper's Thesaurus Linguae Romanae et Britannicae*, otherwise known as *Cooper's Dictionary*.[2] This mighty achievement, admired by no less a redhead than Elizabeth I, who made its author Dean of Christ Church Oxford as reward for his labors, is a building block of the English language and is believed to have been one of the most significant resources used by that great word-smelter William Shakespeare. But the specific chromosome responsible for red hair was only identified in 1995, by Professor Jonathan Rees of the Department of Dermatology at the University of Edinburgh.[3] So for almost the entirety of its 50,000-year existence on this planet, red hair, across every society where it has appeared, has been wrestled with as an unaccountable mystery. In the search for an explanation for it, it has been hailed as a sign of divinity; damned as the awful consequence of breaking one of the oldest sexual taboos; ostracized and persecuted as a marker of religion or race; vilified or celebrated as an indicator of character; and proclaimed as a result of the influence of the stars. It is, unsurprisingly, none of these things, yet at the same time, society's—any society's—responses to red hair have become so inseparable from the thing itself that it has become all of them. And there is as much

---

2 Bishop Cooper's great work almost never saw the light of day at all. When it was half completed, Cooper's wife ("a shrew," according to that indefatigable chronicler John Aubrey), "irreconcileably angrie" with him for neglecting her in favor of his studies, broke into his study and threw his papers on the fire. Aubrey does not record if she was a redhead, too.

3 No one could write a book on redheads without a debt to Professor Rees and his research. I am very happy to record my gratitude to him for his early assistance and generous advice.

mistaken nonsense written about it now as there was a hundred, or two hundred, or, for that matter, five hundred years ago.

Let me illustrate what I mean here by way of a famous red-headed tale, which functions almost as a parable. In 1891 Sir Arthur Conan Doyle published the classic Sherlock Holmes short story "The Adventure of the Red-Headed League." The redhead at the center of the tale is a pawnbroker, Jabez Wilson. Owing to the particular tint of his rare red hair, Jabez is selected by the mysterious League and lured from his pawnbroker's shop to an empty office, where he is paid to spend his time pointlessly copying out chunks of text from the *Encyclopedia Britannica*—a task that, according to the League, he and only he, with his unique flame-red hair, is fit to do (you can see why this wouldn't have worked as a ruse were he garden-variety blond- or dark-haired). Sherlock Holmes, of course, spots at once that the pawnbroker's shop sits next to a bank and that the "work" offered to Wilson by the Red-Headed League is no more than a trick to get him out of the way. The League is simply the cover for a gang of robbers who plan to break into the bank through the pawnshop's basement; and Jabez has been selected not because of his hair but because of the location of his shop. In other words, it is a story whose final explanation is completely different from the one you might expect. In exploring the history of red hair, such will very often prove to be the case.

We live in this extraordinary age in which a butterfly flapping its wings on one side of the planet truly can create a tornado on the other—but only if the butterfly then sets up a website. There is an entire alternate solar system of knowledge and its opposite circling around up there. This is often quite miraculously wonderful—I can fly across the wastes of the Taklamakan Desert in western China, ancient home of the mysterious blond and redheaded Tarim mummies, like Luke Skywalker in his landspeeder, then order up the definitive account of the mummies' discovery by simply dipping my pinkie. What would once have been completely beyond human imagination is now as quotidian as a grocery list, and it seems we are all putting them together. In this universe of information and of *un*formation, some are born redheaded, some achieve it, and some poor souls simply have it thrust upon them. There are endless lists of supposed historical redheads out there, a tangle that links one site to another like binary mycelia; layers of space junk that typify redheads as impulsive, irrational, quick-tempered, passionate, and iconoclastic; great drifting rafts of internet factoids (currently the most notorious: the notion that redheads are facing extinction), with repetition alone creating a kind of false positive, a sort of virtual truth by citation. This, I discovered to my delight, is known as a "woozle," after the mythical and perpetually multiplying beasts hunted in the Hundred Acre Wood by Winnie-the-Pooh and Piglet, way back in 1926. It takes Christopher Robin to point out that the pair are simply following their own ever-increasing footprints in a circle around a tree. This book will endeavor not to add to the population of woozles.

Red is a color that has exceptional resonance for our species. There's an argument that it may have been the first color early primates learned to distinguish, in order to be able to select ripe fruits from unripe, and it still seems to speak to something primal in the human brain today: those suffering temporary color blindness as a result of brain damage are able to perceive red before any other color. And it is full of contradictions. It is the color of love, but also that of war; we see red when furiously angry, yet send our love a red, red rose; it is the color of blood, and can thus symbolize both life *and* death, and in the form of red ochre or other natural pigments scattered over the dead, it has played a part in the funerary rites of civilizations from the Minoans to the Mayans. Our worst sins are scarlet, according to the prophet Isaiah, and it is the color of Satan in much Western art, but it is the color of luck and prosperity in the East. It is universally recognized as the color of warning, in red for danger; it is the color of sex in red-light districts across the planet. The symbolism and associations of red hair embody all these opposites and more.

Red hair has always been seen as "other," but fascinatingly and most unusually, it is a white-skinned other. In the aristocracy of skin, as the historian Noel Ignatiev has described it, and in the Western world of the twenty-first century, discrimination is rarely overtly practiced *against* those with white skin. Yet people still express biases against red hair in language and in attitudes of unthinking mistrust that they would no longer dream of espousing or of exposing if the subject were skin color, or religion, or sexual orientation. And these expressions of prejudice slip under the radar precisely because by and large there is almost no difference in appearance (aside from the hair) between those discriminating against it and those being discriminated against. It is as if in these circumstances, prejudice doesn't count. Attitudes toward red hair are also extraordinarily gendered, something we'll encounter very frequently in the following pages. In brief, red hair in men equals bad, in women equals good, or at least sexually interesting. But even within this simplistic categorization there is a glaring contradiction, since culturally it seems we can get our heads around red-haired men as both psychopathically violent (Viking berserkers; or in the UK, the drunk swaying down the street with a can of super-strength lager in one fist and on his head a comically oversize tartan beret, complete with fuzz of fake ginger hair; or even Animal from *The Muppets*), and as denatured, unmasculine, and wimpish (Napoleon Dynamite, for example, or Rod and Todd Flanders from *The Simpsons*). Stereotypes of redheaded children reflect these opposites, with red hair being used to characterize both the bullied and the bully—Scut Farkus in the movie *A Christmas Story* being a memorable example of the latter—unless the children in question are girls, in which case they are generally perky (Anne Shirley in *Anne of Green Gables*) and plucky (Princess Merida in Disney's *Brave*) and winningly cute (Little Orphan Annie). Redheaded women are supposedly the least desired by their peers of the opposite sex among American college students; yet (and I have to say, this has been my own experience) the popular construct of the female redhead is often profoundly eroticized and escapes the rules and morality applied to the rest of female society.[4] For this far from godly point of view, we have in large part to thank the medieval church.

---

[4] Saul Feinman and George W. Gill, "Sex Differences in Physical Attractiveness Preferences" *The Journal of Social Psychology* 105, no. 1 (June 1978): 43–52.

This happens with red hair, time and again. Its presence, and attitudes toward it—the cultural stereotyping, cultural usage, cultural development—link one historical period, one civilization, to another, sometimes in the most surprising ways and very often flying in the face of all logic and common sense as well. One of the most intriguing aspects of the history of red hair is the way these links run on through time. You begin by investigating the impact of redheaded Thracian slaves in Athens more than 2,000 years ago and end at Ronald McDonald. You examine the workings of recessive characteristics and genetic drift in isolated populations and come to a stop at the Wildlings in *Game of Thrones*. You explore depictions of Mary Magdalene and find yourself at Christina Hendricks.

Why is the Magdalene so often depicted as a redhead? What possible reason can there be for that? If there ever was a specific individual of this name (which is a big assumption—the Magdalene as the Western church created her is a conflation of a number of different Biblical characters), her name suggests that she could have been a native of Magdala, on the western shore of the Sea of Galilee, well below the forty-fifth parallel. Beneath this latitude, redheads, while not unknown, are vanishingly rare, so the Magdalene's coloring in Western art and literature is unlikely to recall some Biblical truth. What, then, for so many artists, from the medieval period onward, is the explanation for showing Mary Magdalene with red hair? What message did that convey to an audience five hundred years ago? What might it tell us about that audience? And what might it illuminate about our own attitudes toward red hair today?

To begin with, from my own experience in my student days of working as an artist's model, I think artists simply enjoy depicting red hair. They like the turning shades and tints, they relish the glint and gleam of light upon it and the way that light bounces off the pale skin that so often goes with it. But the meaning of the red hair of the Magdalene takes one somewhere else altogether. It reflects the fact that the version of Mary Magdalene that the Western church has always found most fascinating is that of a reformed prostitute, a penitent whore, and culturally, for centuries, red hair in women has been linked with carnality and with prostitution. It still is today. What color is the hair of Helena Bonham Carter's character, Red Harrington, in Disney's 2013 film *The Lone Ranger* (and yes, there is a clue in her character's name)? As flaming a red as Piero di Cosimo's poised, calm, intellectual *Magdalene* of *c.* 1500, sat at her window, reading her book. Two entirely different women, centuries apart, yet linked by their societies' identical responses to their hair color. And this leads back to one of the stereotypes that began this discussion, and back to one of the greatest contradictions in the cultural history of the redhead: the centuries-long linkage between red-haired women and sexual desirability, and the fact that (despite, at the time of this writing, the naturally red-haired *Sherlock* actor, Benedict Cumberbatch, being voted sexiest actor on the planet) the exact opposite seems to hold true for redheaded men.[5]

This book is a synoptic overview of red hair and redheaded-ness: scientifically, historically, culturally, and artistically. It will use examples from art, from literature, and, as we come up to the present day, from film and advertising, too. It will discuss red hair not

---

5 *Metro* newspaper, October 3, 2013.

just as a physiological but as a cultural phenomenon, both as it has been in the past and as it is now. Redheaded women and sex and the gendering of red hair is a subject to which it will return in detail, but this book will also journey through the science of redheadedness, its history and the emerging genetic inheritance that is starting to be understood by modern medicine today. It will examine the many conflicting attitudes toward the redhead, male and female, good and bad, West and East. It is a study of other, and as always, what we say about "other" is far less interesting than what that says about us. But if you are going to ask what and who and how and why, the place to start is where and when.

WAY, WAY BACK,
MANY CENTURIES AGO

In solving a problem of this sort,
the grand thing is to be able to reason backwards.

ARTHUR CONAN DOYLE,
A STUDY IN SCARLET, 1887

The ur-redhead, the first carrier among early modern humans of the gene for red hair and thus the genetic grandparent of the vast majority of redheads now alive, appeared on this planet some time around 50,000 years ago.

The world, at this point, was a very different place from how it appears today. Those parts now dry and arid, such as the Sahara, were green and pleasant; areas we think of as temperate, including most of western Europe, were either tundra or under an ice sheet. Stomping or slinking across that ice sheet went the fantastic mega-fauna of the Later Stone Age, or Upper Paleolithic period—woolly mammoth, giant elks, two-hundred-pound hyenas, saber-toothed cats. Trailing after them went Europe's resident population of Neanderthals, who had lived as hunter-gatherers in this landscape for 200,000 years; and creeping cautiously along as a distant and possibly rather puny-looking third came the first early modern humans.

These early humans had left Africa some 10,000 years before. They had already created populations in the Middle East and Central Asia; they were to explore around the coastlines of the Indian subcontinent; reach as far across the Pacific as Australia and as high as Arctic Russia; find toeholds in the Far East; and at some point cross the land bridge into what is now North America. Their expansion was driven (along with, one might suspect, hunger or greed) by an event referred to by paleontologists as the Upper Paleolithic Revolution. What this term encompasses is a step-change in tool-making, from basic stone implements to highly specialized artifacts of bone or flint that range from needles to spearheads; evidence for the first purposeful engagement in fishing; figurative art, such as cave-painting, along with self-adornment and bead-making; long-distance trade or bartering between different communities; game-playing; music; cooking and seasoning food; burial rituals; and in all probability at this date, the emergence of language.

There is no one reason why this evolutionary jump happened when it did, nor even a consensus as to when or where it began. It may have been driven by changes in climate, as the ice sheets receded or grew, causing these early modern humans to create new technologies and survival strategies or perish. It may have been a very gradual process, but simply without a large-enough surviving debris field of earlier artifacts and evidence for us to be able now to judge how gradual; it might have been triggered by some sudden and singular genetic anomaly; or any possibility between these two. We simply don't know; too much evidence has been lost to us. Melting glaciation and rising sea levels have drowned the evidence of the earliest coastal settlements. All trace of those who may have battled out a nomadic existence on the ice sheets has disappeared. What we do know is that some 40,000 to 35,000 years ago, those people who had settled the grasslands of Central Asia began to explore outward, west and north, working their way from Iran to the Black Sea, up the valley of the Danube, and into Russia and the rest of Europe. With them, along with their new technologies and beliefs and emerging ethnicities, they carried the gene for red hair.

This may come as a surprise. The gene for red hair, for pale skin, for freckles (on both of which more later), did not originate in Scotland, nor in Ireland, despite the fact that in both those places you will now find the highest proportion of redheads anywhere on Earth (Scotland leads the way with 13 percent of the population being redheads, and maybe 40 percent carrying the gene for red hair, while 10 percent of the people of Ireland have red hair and up to 46 percent carry the gene). Logically, the place with the greatest incidence of any particular characteristic would, one might think, be the place where it first came into being, but not in the case of red hair. The gene emerged at some point in time between the migration from Africa and the settling of those grasslands of Central Asia.

We're able to make this assertion because of a superbly elegant hypothesis known to scientists as the molecular clock. This piece of evolutionary calculus uses the fossil record and rates of minute molecular change to estimate the length of time since two species diverged, charting this over thousands upon thousands of years, if need be. It enables us to calculate not only how long populations have been separate but how separate they are, with each change in an amino acid, or a DNA sequence, being a tick of the clock. Calibrated to the fossil record, the molecular clock makes it possible to estimate the point in geological history when new genetic traits first came into being. Such as, for example, red hair, or the gene for red hair at least, since at this point that is all we are talking about, the gene, rather than its expression, the appearance of red hair itself.[6] And the reason for this lies in the fact that the number of individuals making these migrations seems to have been quite astonishingly low.

It's been estimated that at the time of that final successful migration out of Africa 60,000 years ago, there may have been no more than 5,000 individuals in the entire African continent. The number estimated to have crossed from Africa over the then-shallow mouth of the Red Sea and into the Middle East, whose footprints now lie fathoms deep and whose descendants would become *Homo sapiens*, may have been no more than 1,000 and could have been as few as 150. There had been other migrations before this one—the fossilized remains of Java Man and Peking Man, or the 1.8-million-year-old remains of *Homo erectus* recently discovered in Georgia represent earlier offshoots. The Neanderthal population of Europe is thought to have descended from another even earlier common ancestor, shared with early modern humans, maybe 300,000 years before. There may have been many other earlier migrations out of Africa that failed—just as there were many Roanokes before there was a single Pilgrim Father—or if they did not, their genetic traces are still too deeply hidden within us for science to be able as yet

6 Accuracy over eons is of course a rather different matter from setting your alarm clock in the morning. Some paleontologists would push the appearance of the gene for red hair back to between 100,000 and 50,000 years ago.

to make them out. But even without ice storms and blizzards of a ferocity that would make headlines in Siberia today, even without enormous carnivores, for this nascent population of early modern humans, the world was not only a staggeringly dangerous place to be, but one with precious few others of the species to share it with.[7]

And how do we know this? Because, despite the many variations in skin, hair, and eye color that loom so large for us, despite the many rich and challenging social and cultural differences across our planet, genetically we are so very *un*diverse.

Consider, for example, the heather, or *Ericaceae*. It colors the hill-sides about my brother's house on the Isle of Skye, fills cranberry bogs in Canada, and romps across the foothills of the Himalayas. Heather, cranberry, and rhododendron are all *Ericaceae*. There are 4,000 different species in total. You need many, many ancestors to create that much genetic divergence. Consider the dog, *Canis lupis familiaris*. There are nine separate breeds of dog, from wolves and jackals to man's best friend, snoozing at your feet. The domesticated breed itself encompasses a spectrum of variation from dachshunds to bulldogs to Great Danes. Think how much genetic diversity that requires. Compared to the differences between a Chihuahua and a St. Bernard, what do we offer? Our limbs are always pretty much in the same proportion to the rest of our bodies, as are our facial features. Our skulls do not radically change shape; we do not come some with

---

7 As an example of both the danger and the isolation faced by our ancestors, the Toba Catastrophe, a supervolcanic eruption that occurred in Indonesia between 77,000 and 69,000 years ago, created a ten-year winter that could have reduced the entire population of the planet to something between 3,000 and 10,000 individuals and may have been one of the catalysts for that ancient migration from Africa in the first place.

snouts and some not; our ears do not stand up or flop, or trail along the ground. Yet we have been breeding and interbreeding for millennia. The reason why there is so little differentiation between any one of us and every other one of us, compared to so many other species of fauna and flora, is because the number of genetically unique individuals the process began with was so mind-bogglingly small.[8] And the one certainty about red hair is that for a red-haired baby to result, both mother and father have to be carrying the gene, and what is more, both have then to donate that specific recessive gene in sperm and egg. So if the gene for red hair was present in the early human population in the same percentage in which it is found today, with such tiny numbers of people spread out over such a huge area, it might have existed unseen, unsuspected, and unexpressed for generation after generation without any such propitious meeting taking place.

It must also be remembered that in discussing our ancestors of all those many millennia ago, almost nothing can be stated as irrefutable fact. There are a half dozen equally valid theories for every tiny piece of ancient evidence. Do we know, for example, that *Homo sapiens* was responsible for the extinction of all those giant animals and the disappearance of the Neanderthals? No, we don't. We know that the arrival of one *coincided* with the disappearance of the other; we know, for example, that the last Neanderthals in Europe, who were physically far stronger and had bigger brains than the incomers who displaced them, had completely disappeared by 24,000 years ago, the last of them dying

---

8 To make matters worse, we are also the only version of our species, *Hominidae*, to exist. No other version of us has survived.

in remote caves facing the sea on the coast of Gibraltar. Looking at our more recent actions on this planet we may conclude that we're a depressingly good candidate for the prime suspect; but the extinction of the cave hyenas and saber-toothed cats (*circa* 11,000 years ago), giant elk (7,000 years ago), the mammoth (the last herd lived on Wrangel Island, off Siberia, as recently as 1,650 years ago), and of the Neanderthals themselves could equally well have been caused by climate change or by disease as by the effects of our aggression or over-predation. Or it might have been caused by a combination of all these things. We simply don't know.

All the same. We are an ambitious, covetous, exploitative, and destructive species, and there has rarely been much good in our engagement with anything we perceive as "other." We interpret difference as threat rather than potential. And sadly, our race is not alone in that.

The El Sidrón caves are in northern Spain, inland from the coastline of the Bay of Biscay. The nearest large town is Oviedo. During the Spanish Civil War, the caves were used as hideouts by Republican fighters. They have always attracted the curious and intrepid, and when in 1994 what appeared to be two human jawbones were discovered in the gravel and mud on the floor of one of the caves, it was assumed that as the remains were in such good condition they were the tragic relics of some misadventure from the conflict of 1936–39—the victims of some forgotten Civil War atrocity.

The bones were indeed evidence of an atrocity, but of one much, much longer ago. The remains of twelve individuals—three men, three women, three teenage boys, and three children, including an infant—were those of an extended family group of Neanderthals who had presumably been ambushed outside the cave, killed, dismembered, and then cannibalized, the flesh removed from their bones with sharp flint knives, the long bones split to get at the marrow. It's presumed they must have strayed into the hunting territory of another, rival group of Neanderthals and paid this dreadful price. Or perhaps conditions were so harsh it was a simple case of them or us. After their deaths, some event, perhaps a storm and flash flood, caused the roof of the cave in which their bones were found to collapse, washing their remains into the caverns, and thus they were immured together for another 50,000 years.

Any discovery of so much material in one place is of course of immense importance to archaeologists, but what makes the El Sidrón family so significant a find is just that: these individuals all seem to have been related to one another. They share no more than three groups of mitochondrial DNA, the type that is passed unchanged from mothers to children. In fact all three of the adult men have the same type. And so well preserved were the remains that forensic science could not only extract readable amounts of fragmentary DNA from them but could even fit individual teeth back into jaws. There are similar idiosyncrasies in dentition between the teeth and jawbones of different members of the group, and two of the men shared the same gene variant, which, it is thought, would have given them pale skin, freckles, and red hair.[9]

9 See http://humanorigins.si.edu/evidence/genetics/ancient-dna-and-neanderthals/neanderthal-genes-red-hair-and-more.

We all possess some DNA (maybe 1–4 percent) in common with Neanderthals, and presumably with that original common ancestor, all those many thousands of years ago. If you reach back in time, the percentage of Neanderthal DNA in modern humans seems to increase. Ötzi the Iceman, whose mummified body, frozen into a glacier, was found in 1991 in the mountains of the Austrian-Italian border and who died in about 3,300 BC, had more Neanderthal DNA than today's modern humans. Not a huge amount (it's been estimated at 5.5 percent), but a statistically significant one.[10]

It would be very neat, therefore, to assume that the gene for red hair as it is found in most redheads today is a Neanderthal characteristic, and that when the two species, Neanderthals and early modern humans, met and interbred (scientific thought goes back and forth on this point, but let's assume usual things happen most usually), the gene was transferred over. It would be very neat; and given the redhead's reputation for violent bad temper, it's a theory that has given amusement and much satisfaction to many non-redheads since the first days of anthropological science back in the nineteenth century. It is still encountered time and again in discussions of red hair today, but it is also completely wrong. The genetic mutation that produced the red-headed men from El Sidrón is different from that found in redheads today. It is instead an example of a phenomenon where what appears to be the same result stems from very different causes—something else we will be encountering again. In any case, the far more fundamental point of interest with the

---

10 See http://johnhawks.net/weblog/reviews/neandertals/neandertal_dna/neandertal-ancestry-iced-2012.html.

red-headed Neanderthals of El Sidrón is not the hair—it's the freckles. It's the skin.

The gene that today results in the red hair of almost every redhead on the planet sits on chromosome 16, and if you have red hair, it's because the version you have of that gene is not working as well as it might. Working perfectly, that MC1R gene, or melanocortin 1 receptor, to give it its full name, would give you brown eyes, dark skin, and an ability to withstand strong sunlight without developing sunburn, sunstroke, or worse. It would do this by stimulating the production of a substance called eumelanin, which colors dark skin, dark eyes, and dark hair. However, MC1R is fritzy. Like a bad internet provider, it flips in and out. If you have red hair, it is almost certainly (there are rare medical conditions that produce red hair, and in the Solomon Islands in the Pacific, an entirely different genetic mutation gives some islanders the most striking gingery-blond afros) *almost* certainly, therefore, down to the fact that you carry two copies of a specific recessive variant of the MC1R gene that dials eumelanin production (with all its protective benefits under strong sun) right down, replacing it with yellow or red phaeomelanin, in an extraordinarily complex set of variants that determines the color of individual hairs on your head and of individual cells in your skin.

MC1R, however, is not alone in this process. There is another gene, HCL2, on chromosome 4 (rather unpoetically, HCL2's full name is simply "hair color 2 [red]"), that also contributes to red hair. Moreover (I did warn you this was complex), while red hair is indeed caused by a recessive gene, there are many possible variants, from recessive red to fully dominant brown or black, with an equal number of manifestations of so-called "codominance" in between.

Thus, many a blonde or brunette has freckles. Many brunettes also have a red tint to their hair, or rather, hair cells that are individually and one by one a mix of eumelanin- and phaeomelanin-producers. A man can have brown or blond hair on his head, yet a beard that grows in red (and very annoying some of them seem to find it, too). This pairing of, say, brown hair, freckles, and red beard is known as "mosaicism," which describes it pretty much perfectly. You can have red hair in any shade from palest strawberry blond to deepest chestnut. It can come hand in hand, as it were, with blue/green eyes and pale skin and freckles, as it has with me, or with amber eyes, or hazel, or dark brown. A recent survey undertaken by Redhead Days, who run the largest annual festival of redheads in the world, uncovered the fact that up to a third of those who responded classified their eye color as hazel or brown. And (and this is where things get really interesting) it can come with skin dark enough to protect you from sun that would send many another redhead running for sun hat, sunglasses, and sunblock. In general, redheads and strong sun most emphatically do not mix. If you look back at the Redhead Map of Europe, at the beginning of this book, you will see how suddenly the incidence of red hair drops off below that forty-fifth line of latitude. But Afghanistan, Morocco, Algeria, Iran, northern India and Pakistan, and the province of Xinjiang in China all have ancient, native populations of redheads. Shah Ismail I (1487–1524), commander in chief of the Qizilbash and founder of the Safavid dynasty in Iran, was described by the sixteenth-century Italian chronicler Giosafat Barbaro as having reddish hair, and indeed his portrait in the Uffizi, by an unknown Venetian artist, shows a man with an aquiline nose, red beard, and red mustache. The sixteenth-century *Shahnama* of his Safavid successor, Shah Tamasp, now in the Metropolitan Museum of Art, shows a red-bearded hero, Rustam. In the ancient world Alexander the Great's Roxana, who was born in Bactria, now northern Afghanistan, and who the Napoleon of the ancient world married in 327 BC, was reputedly a redhead; the present-day Princess Lalla Salma of Morocco perhaps gives an idea of quite how beautiful she might have been. Then there is that different shuffle of the genetic pack in the Solomon Islands in the Pacific—which, again, comes with skin dark enough to withstand in this case tropical sun. (Fig. 29). In the history of red hair, what you might call its typical Celtic manifestation—pale skinned, blue eyed, suited for cool and rainy climes and cloudy skies—may not originally have been typical at all. In fact pale skin in early modern humans has been estimated as having appeared as recently as 20,000 years ago.[11]

Most genetic mutations die out rather than become more common, and if they are as unadvantageous as pale skin under fiercely sunny African skies, they take their unfortunate carriers with them.

---

11 Other studies give the date as being between 12,000 and 6,000 years ago. One estimate is as recent as 100 generations, or a mere 2,500 years ago, although this is rather unconvincing. Had pale skin suddenly begun appearing among the native populations of the countries they had conquered, surely one of those indefatigable chroniclers of the Roman world (many of whom we will meet in the next chapter) would have made specific mention of it. Thanks to my colleague Kate Owen for this point.

Blue eye color, meanwhile, or rather the switching-off of the gene for brown eye color (although the genetics of eye color are so astonishingly complex as to admit almost any possibility) is thought to have come into being in what is now modern Romania, perhaps 18,000, perhaps 10,000 to 6,000, years ago.

It may be going too far to characterize Mother Nature as smart, but left to herself she does have a heartlessly effective way of cleaning out the gene pool. If you're not fit to survive, you won't. But if a genetic quirk confers a benefit upon those carrying that gene, it, and they, will flourish. And pale skin under Northern skies does exactly that. If your eumelanin production is dialed back, if you have pale skin rather than dark, your body will be much more effective at synthesizing vitamin D, using whatever sunlight is available, than if your skin were darker. And as the ice sheets retreated, and that growing population of early modern humans moved from Russia into Scandinavia, and eventually into the whole of northern Europe, the absence of strong sun in these climes allowed the MC1R gene to mutate into what geneticists term "dysfunctional variants" without these variants proving fatal to those carrying them. In fact the farther north this population went, the more advantageous pale skin became. If you have enough vitamin D, your skeleton develops as it should. If you do not, your bones are soft and stunted, and as you learn to walk, your legs bow under your weight. This is osteomalacia, or rickets. In adults, it causes calcium to leach away from your bones; in children, it cripples you. Women of child-bearing age who suffered from rickets when they themselves were children have distorted pelvises that make carrying a pregnancy to full term difficult and childbirth hazardous, if not fatal (Fig. 3).

Communities that eat most of their protein in the form of meat, as with the early hunter-gatherers, rarely suffer from vitamin D deficiency. But the tendency in the early modern human population was to settle; to become farmers, to grow and eat grain. In these circumstances, pale skin helped keep you strong and healthy. In

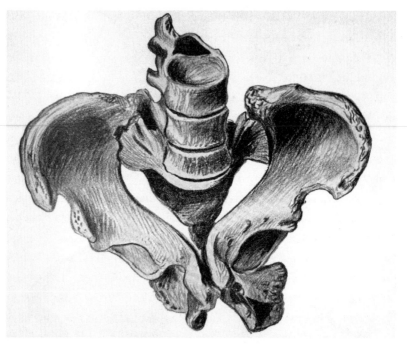

Fig. 3 Medical illustration of the pelvis of a woman suffering from rickets, showing the narrowed and distorted birth canal.

particular, it gave a significant advantage to women during pregnancy and breast-feeding, when their bodies' demand for vitamin D was at an all-time high, which along with all the other ancient and instinctive associations of the color red (fire, blood, passion, ripeness) does rather open the question as to whether the often highly sexualized image of female redheads might not start here, with the simple fact that choosing a redhead as a mate meant you bred successfully, and that your pale-skinned children, themselves now carrying the gene for red hair, did the same.[12]

This is also where the random mysteries of "genetic drift" come into play. Genetic drift is the term used to describe random changes in the frequency of a genetic variant, or "allele," as it is also known within a given population. There is some essential scientific terminology to become familiar with here, and perhaps the simplest way to conceptualize it is if you cast your mind back to your first school photograph. You, seated there cross-legged on the ground, squinting at the sun, are an individual allele, an individual genetic variant. The front row of you and all your classmates make what is known as a haplotype—a cluster of linked alleles, all of whom are likely to be inherited together. Your haplogroup would be everyone in the school.

If a population is large, it can take a considerable time before random changes in the frequency of an allele become noticeable, and indeed as they are governed only by chance, and their frequency can alter one way or the other from one generation to the next, some may never result in any noticeable change at all. But in a small population, the fixation of a particular allele within that group, such as (for example) red hair, can happen very rapidly indeed. In small populations, in borderlands, in any community set apart from the great genetic ebb and flow of the human ocean, in places such as Ireland or on the west coast of Scotland, its effect can be established within just a few generations. Add to this the not-random-at-all mysteries

---

12 In light of the association between red hair and successful childbirth, it's intriguing that C. G. Leland records in his *Gypsy Sorcery and Fortune Telling* (1891) the belief that for an easy birth, red hair should be sewn into a bag and worn next to the skin of the belly during pregnancy.

of sexual selection, and the fixing of red hair among these liminal populations, and among liminal populations in the Levant, the Caucasus, and the Atlas Mountains, becomes a phenomenon that exists logically and obviously. As of course it also does among the famous Wildlings in *Game of Thrones*. Redheads: guaranteed throughout history to crop up in the last place you expect them.

BLACK AND WHITE
AND RED ALL OVER

Men make gods in their own image; those of
the Ethiopians are black and snub-nosed, those
of the Thracians have blue eyes and red hair.

XENOPHANES

The house of Marcus Fabius Rufus, on the so-called Vico del Farmacista, is the largest domestic dwelling so far excavated in Pompeii. It has four separate levels and appears to have been continuously occupied from the time of the Roman conquest of the then–port of Pompeii, in 80 BC, up to the destruction of the city in AD 79. M. Fabius Rufus (whose body may very well be one of the four found within the house, and whose name was preserved for us in a piece of scurrilous below-stairs graffiti) was but its final owner; over its 160-year history the house had many others, all of whom contributed in some way to the layout and decoration of its rooms—opening up doorways here, closing them there, enlarging this room, repainting that. Then came the eruption of Vesuvius, with its earthquakes and choking blizzards of ash; and when archaeologists finally opened up the house of M. Fabius Rufus, they found the earthquakes of the eruption had partially demolished a wall in Room 71 and revealed behind it another, the style of whose fresco decoration dates it to the century before.

The fresco shows a woman standing between two partially opened doors. She has large eyes, a delicately rounded mouth, and a fashionably piled-up hairstyle; she holds a child against her shoulder (he's maybe a year old, an elongated Roman *putto*, held against his mother with his back to us and his naked rump bared endearingly to posterity); she wears what has been interpreted as a royal diadem, and her hair is a warm, or one might say a reddish, brown. Recent research suggests that the woman is intended to be Cleopatra, in the guise of Venus Genetrix, the mother-goddess of the Roman world, and that the child should be identified as her son Caesarion, born in 47 BC, whose father, so his mother claimed, was Julius Caesar.[13]

The identification of the woman is based upon her likeness to two marble busts of Cleopatra, one now in the Vatican (which may also once have been completed by the figure of a child), the other in Berlin; the presence of the diadem; and the fact that the two half-open doors and the rest of the frescoed scene around her seem to allude to the appearance of the temple to Venus Genetrix, set up by Caesar in the Forum Julium in Rome in 46 BC. The temple was graced with a gilded statue of the goddess, widely and scandalously reputed to have been modeled on Cleopatra herself, who was in Rome from 46 to 44 BC. The thinking is that on Caesar's assassination, in 44 BC, and the accession of his official heir, his great nephew Octavian, as the Emperor Augustus, the then-owner of the house hid the

13 Susan Walker, "Cleopatra in Pompeii?," *Papers of the British School at Rome* 76, (2008): 35–46.

fresco behind a wall to cover up his own Julian sympathies.[14] Thus, intriguingly, the history of red hair and the world of Roman real-politik cross paths. But is any of this proof, as has been claimed, that Cleopatra was a redhead?

Or to put it another way, *what*, and *who*, is red? One person's unmistakable red is another's vaguely chestnut, and even that authoritative-sounding quote from Xenophanes is not as undisputable as it appears: Xenophanes wrote in ancient Greek, and some authorities translate his "red" as "fair." Moreover, his writings have come down to us only via their use in the writings of others. The quote was preserved, five centuries after Xenophanes (*c.* 570–*c.* 475 BC) worded the original thought, in the work of Clement of Alexandria, a founding father of the Christian church. That so-called portrait of Cleopatra in the house of M. Fabius Rufus is the result of a similar process of transmutation. Judging by her official likeness on the coinage of her rule, Cleopatra herself had neither large eyes nor a rounded mouth. The coins show instead a woman with a long nose, almost hooked, and a sharp, knowing smile. Also, Cleopatra was a native of Egypt, obviously enough, a country that lies well below the forty-fifth parallel, which makes the possibility of her being a natural redhead unlikely. Then one has to remember the Egyptians used wigs, some of which, recovered by archaeologists, have proved to be made from the reddish fibers of the date palm. They also dyed their hair. In fact they seem to have applied as many colors, gels, and waxes to their hair as we use today. The mummy now in the Egyptian Museum in Cairo of the great

---

14 This would have been only smart. Caesarion himself was killed, probably by strangulation, on Augustus's orders in 30 BC at the age of seventeen—eleven days after his mother's famous suicide.

pharaoh Ramesses II, who ruled Egypt some 1,200 years before Cleopatra, has dyed red hair. Ramesses died in 1213 BC, at the age of ninety, and his own hair by this point was white, unsurprisingly, but either before death, or as part of the embalming process it was dyed with henna.[15] One might speculate that this emulated its color in life, but this would be speculation only. In fact red hair seems to have had something of a conflicted history in Egyptian culture. Red was the symbolic color of Set, the god of violence and disorder and lord of the hostile desert (there is supposedly an Egyptian prayer to Isis, begging for deliverance from "all things evil and red"[16]). Before Ramesses's family, who were followers of Set, came to power, supposedly every year a redheaded male was burned alive as a sacrifice, or so the Roman chronicler Diodorus Siculus, writing more than a millennium later between 60 and 30 BC, informs us. These commentators and chroniclers of the ancient world are the only speaking witnesses we have, but with all of them, their words come to us as light from a dead star, distorted as echoes, sometimes as little more than static, down a long, long line from far away. We cannot know all of the details or sometimes indeed any of the circumstances in which they were writing; we can only guess, and to use these sources to reconstruct that ancient world is to be a detective—almost an archaeologist—yourself. None of this is as black and white as it may first appear.

---

15 For the arguments that red was Ramesses's hair color in life, see http://www.lorealdiscovery.com, and Bob Brier, *Egyptian Mummies: Unraveling the Secrets of an Ancient Art* (New York: William Morrow & Co., 1994), 153. The question is still open, however: recent research by Silvana Tridico suggests that hair decays after death and its color can alter as a result of fungal or bacterial growth upon it. See http://rspb.royalsocietypublishing.org/content/281/1796/20141755.

16 Quoted in Eva Heller, *Psychologie de la couleur: Effets et symboliques* (Paris: Editions Pyramyd, 2009) 45.

But back to Cleopatra. What the fresco in Pompeii records, therefore, is not the appearance of an actual woman but the appearance of a statue. It is a sizable and significant step away from the living, breathing, original. Rome was 150 miles from Pompeii, not too far a journey, by any means, for a painter on a commission. Statues in the ancient world were rarely left in that blanched state of seashell whiteness in which we see them today. Once carved, they were often painted, and the grandest were gilded, too. So the coloring of the figure in the fresco may record the original coloring of the now-lost statue, or it may simply bear testimony to the paints available to the painter. The palette of the ancient world was limited to earth colors, mineral pigments, or vegetable dyes, but no matter what the painter intended, the iconic force of his subject would still overpower its actual appearance. The eye of the beholder is all. With so many of those names on website lists of historic redheads, the one thing that they all have in common (aside from the fact that we are most unlikely ever to be able to reach an indisputable conclusion as to their hair color) is that they all behaved as if their hair *should* have been red. We look at this Pompeian fresco, and we see not a woman with reddish-brown hair, we see Cleopatra, Caesar's mistress, lover of Mark Antony, the queen who hazarded a kingdom and chose death over conquest, an archetype to which this book will return over and over again: the flame-haired seductress, exotic, sensual, impulsive, passionate. We see her hair as red because we want to do so. What other color would it be?

So who were the redheads of the ancient world? How real, or not, were they?

The kingdom of Thrace, if such a set of tribal territories can be so referred to, existed from roughly 1000 BC to the final dissipation of the Roman Empire some 1,700 years later. It sat across the western side of the Black Sea and stretched down to the Aegean, over an area that now includes most of Bulgaria and parts of Turkey and Greece. The Thracians were horsemen and warriors, and early contact with them, as both Greece and later Rome would discover, tended to be very bloody indeed. Even their war dances were violent enough to leave the odd participant lifeless. According to the Greek historian Herodotus, the Thracians believed that "to live by war and plunder is of all things the most glorious."[17] When as a Thracian you weren't occupying yourself with that, you might perhaps be indulging in a popular drinking game, consisting of standing on a rock, with your head in a noose. One of your friends kicked the rock away and then the trick was to be quick enough with your Thracian short-sword to slice through the rope before you throttled—a sort of Thracian Russian roulette.[18]

The Greeks were recruiting Thracian mercenaries into their armies as early as 600 BC, by which time they had a strip settlement of Greek trading posts along the Thracian coasts. Alexander the Great, three hundred years later, would do the same, both fighting the Thracians and signing them up. To this day, a common explana-

---

17 Herodotus, *The Histories*. The full text is available on Project Gutenberg. See https://www.gutenberg.org.

18 Lionel Casson, "The Thracians," *The Metropolitan Museum of Art Bulletin* XXXV, no. 1 (Summer 1977) 3–6.

tion for the unexpected appearance of green eyes or red hair in a child in Afghanistan or Kashmir is the onetime presence of Alexander's troops in those regions more than two thousand years ago, and who is to say such hand-me-down folklore doesn't still preserve some kernel of genetic truth? The Thracians were as prized as troops as they were feared as enemies. In 73 BC Rome found itself facing an internal revolt led by the gladiator Spartacus, a Thracian from the border tribe of the Maedi with a Roman military background (the Roman chronicler Plutarch gives us a wealth of this familial detail on him but, maddeningly, fails to record the color of his hair). In addition to their military skills, the Thracians were superb metalworkers in bronze and gold, and they also had a highly evolved belief system covering the underworld and afterlife, as evinced by their elaborately decorated tombs. They lent much of this mythology to the Greeks in turn, although the gods of Thrace seem to have been even darker and less tractable than those of Olympus. The Thracians were also, notoriously, within the ancient world, "barbarian." Not only did they disdain speaking Greek, they also refused to give up the traditional structure of their society, rejecting the whole idea of creating and living in cities, and remained in their small tribal communities. "If they had one head or were agreed among themselves," Herodotus observes sadly, "it is my belief that . . . they would far surpass all other nations. But such union is impossible for them."

To the north of Thrace, at the top of the Black Sea, were the lands of the Scythians, equestrian tribespeople with an origin as far back east as Iran. There is also Biblical mention of the Scythians, in Colossians 3:11: "Here there is no Gentile or Jew, circumcised or uncircumcised, barbarian, Scythian, slave or free," which sounds as if the Scythians are being used as an even more extreme example of barbarism. They too were noted warriors, feared as archers in particular, and flourished from around the seventh century BC to the fourth century AD. Not losing sight of the fact that ancient writers used the term "Scythian" pretty broadly (as, perforce, do archaeologists today), Herodotus is our guide here as well. Writing of a city, Gelonus, in the northern part of Scythia, he describes its people, the Budini, as "a large and powerful nation: they have all deep blue eyes [or gray, depending upon translation] and bright red hair." It's thought that the Scythians, specifically the Budini, might have been the ancestors of the Udmurts of the republic of Udmurtia in Russia, on the Volga River.[19] Why is this thought? Because ever since the anthropologists of the nineteenth century encountered them, the Udmurts have been celebrated as among the most redheaded people on earth, with almost as high a percentage of redheads in their population as the Irish and the Scots. They still are. The Udmurts and the area around Udmurtia on the Volga are the hotspot on the map of European redheads, north of the Caspian Sea.

This is intriguing enough. Even more so is the possibility that the backstory of the Scythians, as it were, might reach even farther eastward, as far as Tibet, Mongolia, and the border of modern China, and that the ancestors of the Scythians might be linked with the civilization of the Tarim Basin, and the Tarim mummies.

The history of red hair is tied to the history of human migration, of one people encountering another, or even "an other," and each of

---

19 The site of Gelonus has been sought by archaeologists for some time; various ancient settlements have been put forward as candidates in the Ukraine or along the Volga River.

these encounters adds another layer to the cultural response to red hair, right down to the present day. Red hair is an unmistakable and very convenient marker of these encounters and is tied in particular to four great human diasporas. Those of the Celts, the Vikings, and the Jews are to come. But let's begin with the first of these four key diasporas: that of the tribes who made the journey across the Middle East to the shores of the Black Sea, and then settled the valley of the Danube (which may itself have acquired its name from a Scythian loanword). If you were to stretch the history of these people eastward, rather than west, and back in time, back beyond the Thracians and Scythians, back even beyond the reign of Ramesses the Great, you would reach the grasslands of central Asia where, it is believed, the history of red hair began. And if, thousands of years ago, rather than trekking west, you had turned east, you would eventually have reached what is now the Taklamakan Desert, in the Tarim Basin.

Almost everything about the civilization of the Tarim Basin and its discovery by Western archaeologists sounds as if it should have come straight out of *Indiana Jones*. There are the stories of the first European explorers who reached the area, to begin with: Nikolai Przhevalsky, who gave his name to Przewalski's Horse, and whom internet mythology would have as the father of Joseph Stalin; or Albert von Le Coq, a German beer and wine magnate who began studying archaeology at the age of forty, whose expeditions were financed by none other than the German emperor Wilhelm II, and who ended by shipping more than seven hundredweight of artifacts back to Berlin, convinced that the presence of red-haired, blue-eyed figures in the frescoes he chipped, carved, and sawed out of caves in northwest China meant he had discovered a new Aryan heartland;

or Sir Aurel Stein, who owed his knighthood as much to the role he played as a spy in the "Great Game," the battling-out between Britain and Russia for influence over Central Asia, as he did to his archaeological discoveries. These early archaeologists discovered the ruins of settlements, orchards, and oases that had once been shaded by poplar and tamarisk trees and watered by rivers that had run dry centuries before, all now buried under the sand dunes of the Taklamakan Desert (the name can be translated as "You go in, but you don't come out"). And, at various graveyard sites around the rim of the Tarim Basin, they also discovered the Tarim mummies themselves, hundreds of them at least, almost perfectly preserved by the cold, dry climate, a climate so perfectly suited to mummification that it has preserved even the bodies of ancient mice in the remains of ancient granaries. What these Tarim mummies reveal is the fact that, in what is now western China, in the province of Xinjiang, bordered on one side by the 'Stans, and on the other by Mongolia, from at least 2000 BC to roughly AD 200 there lived a people of almost modern height with fair skin and blond hair, and in a couple of cases at least, as reported in the authoritative history of these discoveries, with actual red hair. They had angular, Caucasian features and light-colored eyes set in very un-Asian recessed eye sockets. And, they wove, wore, and maybe traded textiles that link them to the tribes of Celtic Europe.[20] Basically, 4,000 years ago, there were people living in western China who looked as European as the tribes living

---

[20] The magisterial study on the Tarim mummies is by J. P. Mallory and Victor Mair, *The Tarim Mummies* (London: Thames and Hudson, 2008). Their hair color is variously described as blond, fair, or red, but as far as I am aware, none of the mummies has as yet been specifically tested to see if they carry the MC1R gene.

at that time around the Seine or the Thames. The Roman writer Pliny the Elder (AD 23-78) who, like M. Fabius Rufus, also perished in the eruption of Vesuvius (Pliny the Elder was both corpulent and asthmatic, and couldn't make it to safety through the pumice fall), included a description in his *Natural History* of these people of western China, given to him in turn by a diplomat from Taprobane (modern-day Sri Lanka), who was visiting the Emperor Claudius:

> These people, they said, exceeded the ordinary human height, had flaxen hair, and blue eyes, and made an uncouth sort of noise by way of talking, having no language of their own for the purpose of communicating their thoughts.

The "uncouth sort of noise" may have been a very early Indo-European language, which scholars now refer to as Tocharian. The people speaking it are given the name "Seres" in Pliny's account, or "people of the land of silk." This is a very big clue as to how and why contact might have existed between the people of the Tarim Basin and those of the Black Sea and even farther into Europe: Tarim was on the Silk Road, one of the most important global trading routes that has ever existed.

We think of our planet as being divided into continents and countries, each of which is the place of origin for people of a specific appearance: African, Eurasian, Caucasian, and so on, but this is simply how the world looks to us at our moment in time. Thousands of years ago, these boundaries might not have been the same. Our ancestors were intrepid and tireless explorers; it may have been 2,500 miles from Cornwall to the shores of Phoenicia, but trade existed between the two. It may have taken seven months to travel overland from the Hindu Kush, between Pakistan and Afghanistan, to China, but the journey was made. (Clearly, so was that—more than 7,500 miles—between Sri Lanka and Rome.) And if goods were being traded, it would be very strange if tribal alliances and marriages weren't taking place as well.

The Scythians left tomb-mounds, known as *kurgans*, across the whole of Eurasia, from the Ukraine to the Altai mountain plateaus between Mongolia and Siberia. Many of these have proved to be rich in the most exquisite examples of metalwork, in gold and bronze. They have also yielded up human remains, which in turn have revealed that the Scythians belonged to a haplogroup, R-M17, that is much more closely related to people living now in eastern Europe than it is to those of central Asia, and which would have given them fair skin, blue or green eyes, and light-colored hair.[21] Hence the link suggested between the ancestors of the Scythians and the ancestors of the fair-skinned, Caucasian-featured mummies from Tarim. This also raises at least the possibility that those unexpected red-haired, green-eyed children of Kashmir and Afghanistan record not the later passing of Alexander's troops in the 320s BC but are perhaps reminders of an even more ancient lineage of some of the earliest redheads on the planet. What the tombs of the mobile, nomadic Scythians, who kept their art small and portable, have not so far rendered up is much in the way of paintings that shows us how these people depicted themselves. The tombs of the Thracians, however, have.

---

21 Some of these remains were of women, but buried with the accoutrements of warriors, leading to the suggestion that Greek stories of Amazons came from their neighbors the Scythians.

Some hundred miles east of Sofia, almost exactly in the center of modern Bulgaria, there is a valley containing no fewer than three hundred Thracian tumuli, or tomb-mounds. One of these is the so-called Ostrusha tomb, dating to 330–310 BC. Like most such tombs, its funerary bed is now empty (the reputation of the Thracians as metalworkers and the lure of Thracian gold meant many tombs were robbed from a very early date), although in this case its original occupant might not have been buried alone. When the tomb was opened, the skeleton of a horse was found inside it, with a knife rusting on its chest, suggesting that the poor beast was led in there and killed by being stabbed through the heart, in order to accompany its master to the Thracian heaven depicted above the bed, on the ceiling.[22]

The coffered ceiling of the Ostrusha tomb is extraordinary. Carved out of solid rock, it is divided into square fields, deeply inset, with both the framing borders and the central square being appropriately decorated. The coffers are painted with scenes of mourning (one, for example, shows the goddess Thetis mourning her son the Trojan hero Achilles—also a redhead, according to some accounts) and of the journey into the afterlife, and in coffer 32, there is the head and shoulders of a young woman [Fig. 4]. Her head is tilted to

---

22 Julia Valeva, *The Painted Coffers of the Ostrusha Tomb* (Sofia, Bulgaria: Bulgarski Houdozhnik, 2005). With grateful thanks to the author for providing me much information and a copy of her superbly detailed book.

the left, as if looking down on the funerary bed, and she is extremely fair of face, even with the damage 2,300 years have inflicted on the fresco. She has skin like a rose petal, an air of gentle, clear-eyed calm that still catches the heart, and she has red hair. It is possible to conjecture that in this setting she represents Demeter or her daughter, Persephone. Both goddesses had powers within the cycle of life, death, and rebirth and were intimately connected with the notion of the turning of the year, and it is natural to see the young woman's red hair as symbolic of fire, of the low winter sun, of sunsets and sunrise, of returning life, of winter and spring.

One hundred miles or so farther east of Ostrusha, toward the coast of the Black Sea and close by the village of Alexandrovo, is another tomb. This too dates to the fourth century BC. Its configuration differs from that in Ostrusha—there is no coffered ceiling; instead in cross-section it resembles an igloo, with a tumulus of earth mounded up over it. You enter through a low, narrow tunnel, as did the treasure hunters who discovered the site as recently as December 2000, and in the main chamber of the tomb, as you stand upright, you see above your head the frescoed decoration of a whole year's worth of hunting scenes. The quarry are boars and deer, depicted as the huntsman thrusts his spear down their gullet or the hounds leap upon their backs. The hunters are shown wearing short tunics, or warmer trousers and boots (one reason for seeing these scenes as taking place in different seasons).[23] One hunter is on

23 G. Kitov, "New Discoveries in the Thracian Tomb with Frescoes by Alexandrovo No 1," *Archaeologia Bulgarica*, 9 (2005): 15-28. One should note that even after his death in 2008 Georgi Kitov is a controversial figure among the archaeologists and scholars of Thrace. A reconstruction of the tomb and its frescoes can be seen at http://www.aleksandrovo.com/en.

horseback, and it has been conjectured that he is the "hero" figure, who might have been buried here. In that case, who are the others? One scene shows a chubby man with a prominent belly, stark naked, enraged and armed with an axe, charging toward a deer twice his size. Another hunter, heavier muscled—he has been described as soldierly—stands with spear raised, ready to deliver the coup de grace to a boar. This hunter's legs are bare and might be interpreted as sunburned; his hair is dark, but when it came to the beard, the artist deliberately changed his palette and painted the beard red.

You can go badly astray in trying to read the ancient world as if it were our own. Archaeologically, it is clear that the Alexandrovo tomb must have been opened at least twice; it also contains the remains of what might have been a stone couch, or might have been a table. The decreasing height of the passage into the central chamber, with the frescoed hunt eternally circling its ceiling, might suggest that the chamber itself was used as a temple, and it and the couch/table inside it were for ceremonies and rituals lost to us. But the scenes and details are so specific—the portly naked man in such furious pursuit of that deer, and the red-bearded "soldier"—that they tempt one to another explanation: that they and the man on horseback hunted together, that they would have recognized themselves and the events in these scenes, and that on the rider's death, for a period, his friends gathered here, and drank, and feasted, and remembered. And this is what they looked like. And one of them had a red beard.

Herodotus is the man to provide some context here. Herodotus was born *c.* 484 BC (so a little after Xenophanes's long life came to its end), in what is now Bodrum in southern Turkey. Thrace would have been no more remote and its customs no less known to him than those of Canada are to the United States. Around 450 BC Herodotus was in Thrace, and in his *Histories* he describes the various habits of the various Thracian tribes. Some, he says, practiced polygamy, with the most favored wife following her husband to the grave and, indeed, fighting for the privilege of doing so. He details with relish how the Thracians kept "no watch" on their maidens "but leave them altogether free," and how unwanted children— possibly, one imagines, the resulting unwanted children—would be sold to slave traders. He also describes their funeral rites and their worship of Dionysus. Thrace was reputedly the birthplace of Orpheus, the musician whose songs were so sweet they could divert rivers, coax trees to dance, and almost freed his dead wife, Eurydice, from the underworld. It was also the site of his death, torn limb from limb by Thracian women supposedly in the throes of Dionysian ecstasy.

You can also, of course, go wildly astray by taking the words of Herodotus at face value, as was noted by writers from Plutarch to Voltaire and as is still debated today. Where Thrace is concerned, though, he seems to have known what he was writing about. And for the Greeks, with their city-states and hierarchy of gods and government, and in particular the Athenians, with their strict notions of order and their rigid seclusion of women to the home, the Orpheus story must have been the perfect Thracian myth. It is inescapably violent and full of both transgressive females and the terrifying mysteries of the afterlife. Thrace both appalled and entranced the Athenians. And it was their single greatest source of slaves.

Besides their depictions of themselves in their tombs, the other means by which we know what the Thracians looked like is from their appearance in Greek art. Thracian women (when not pursuing Orpheus) shuffle slipshod and disconsolate around many a Greek vase. Their hair is shorn short, and their limbs are decorated with tattoos. To the Greeks, who punished escaped slaves with tattoos or branding, this was a sure sign of servitude, but to the Thracians themselves, ironically, these rosettes and dotted lines, these whorls and stylized animals (twiggy-antlered stags seem to have been a particular favorite) were signs of noble birth.[24] Thracian men are depicted as warriors, sometimes fallen, sometimes not. They sport pointed beards, they wear cloaks decorated with bands of geometric patterns and caps that Herodotus informs us were made of fox-skin. There is obvious potential here for confusion between the color of a fox-skin cap and that of the hair of the head, similar to the confusion engendered between the Qizilbash warriors of thirteenth-century Anatolia and *their* crimson headwear. In the case of the Thracians, it may also reflect a connection made in the ancient world between the behavior of the animal—supposedly cunning, sly and untrustworthy—and the character ascribed to redheaded barbarians of whatever tribal identity. As the third-century *Physiognomonica* would have us believe, "The reddish are of bad character.

---

24 Bodies found in Scythian *kurgans* also sport elaborate tattoos, for example that of the famous Siberian Ice Maiden, discovered in 1993. See the article by Anna Liesowska in *The Siberian Times*, October 14, 2014: http://siberiantimes.com/culture/others/features/siberian-princess-reveals-her-2500-year-old-tattoos/.

Witness the foxes." (Of course now it is used of redheaded women in the sense of the 1967 Hendrix song "Foxy Lady," and is thus part of another of those historical associations of red hair with a shelf life of millennia.) But the Thracians are also sometimes shown, unmistakably, as redheads, as with King Rhesos of Thrace, a supporter of the Trojans in the *Iliad*. King Rhesos was late getting to Troy (a little local trouble with those pesky Scythians); then before he had so much as set foot on the battlefield, he was done to death in his tent by Diomedes and Odysseus, who stole his famous horses, too. His death is shown on a black-figure vase now in the Getty Museum in Los Angeles (Fig. 5). Rather less elevated is a little terracotta figure of a runaway slave from the British Museum in London (Fig. 6). It's thought to have been made in Athens, at about 350–325 BC, so the same century as the Alexandrovo tomb. It's no more than five inches, or about thirteen centimeters, high, and it shows a chubby little man, this time sitting on an altar. His shoulders slouch, his left hand grasps his left knee, and his right is raised to his ear, as if he is hard of hearing, or perhaps (given the open wail of his mouth), someone has just clipped him around the head, and here he is, scrambled atop the altar, claiming sanctuary, wailing and bemoaning his lot.

These little terracotta sculptures were traded all across the Classical world, from North Africa to southern Russia. They were the Toby-jugs (or maybe today the bobbleheads) of ancient Greece: cheap, coarse, and dispensable. Scenes of tragedy, such as the death of King Rhesos, were reserved for pottery of the grander sort—the black-figure vases, for example, which were intended for those who could appreciate them. Figures such as this runaway slave derived from comedy and were intended to appeal to the common sort.

And they reflect exactly the "Three Stooges" slapstick aesthetic of Greek comic drama. On stage, our runaway slave would have had a red leather phallus bouncing between his legs under his miniskirt-length tunic (it makes this little figure even more poignant, that his manhood has been broken off). The grotesque expression of the mask he wears would have spoken to his audience of the primitive and ungovernable nature of his emotions. And his hair, or rather the wig attached to his mask (and even now on the terracotta, traces of pigment remain), his hair would have been red.

The Greeks liked lists. They liked the world ordered and subdivided. In the second century AD, the Greek scholar and rhetorician Julius Pollux compiled one of the world's first thesauruses: a dictionary arranged not alphabetically but by subject matter. Within this, the *Onomasticon*, he includes descriptions of the seven different slave types in Greek drama, and four of the seven have red hair. On stage, the names of their characters repeat the same message: Pyrrhias, a slave in Menander's comedy *Dyskolos*, ("The Grouch"), whose name means "Fiery." In the comedies of Aristophanes no fewer than five different slave characters in five separate dramas share the same name: Xanthias, meaning "Goldy," or "Red."

It seems unlikely, to say the least, that every slave in every Greek household was a Thracian redhead. Red hair is still recessive; it would have been at least as much of a rarity in the ancient world as it is today. But in a phenomenon that repeats throughout history, this clearly made no difference. Where "other" is concerned, we focus upon its epitome to the exclusion of every other detail. One thing comes to stand as a symbol for all, and this one characteristic, red hair, came to stand for an entire class, if not in fact an entire nationality.

The association was created; at some subconscious societal level it was accepted, and it stuck. Red hair, in the ancient world, equaled "barbarian"; then via the Greek stage and figures such as our little terracotta runaway, it began to equal "clown." You can trace a line of development from these primitive slave characters of the Greek stage to the white-faced, red-haired clown of the circus big top to Ronald McDonald, unnerving children across the planet (and originally incarnated by Buttons, Ringling Brothers' red-wigged clown); to the rather more endearing Obelix, Asterix the Gaul's red-pigtailed bosom buddy in the cartoon books by Goscinny and Uderzo. It is as if we are watching two redhead archetypes, the ungovernable savage and the comic buffoon, coming into being before our eyes. In fact, it is not even as if. We *are*.

All freeborn people, as the historian Sandra Joshel puts it, are defined by their physical integrity. So to insult their appearance is to insult both their social standing and personal identity. But slaves had no personal identity. They were identified only by the work they might be fit for, like so many differently sized tools in a tool kit: this one has a singing voice. This one would make a good plowman. This one could be a wet nurse. This one could work with vines. And they had no ethnicity either, so red hair no longer marked you out as Thracian; it marked you out instead as disempowered, subservient. The good slave was one who accepted this and subsumed their own identity; the "bad" slave, and the one who got the most laughs on stage, was the one who insisted on retaining a human personality— always a bad one.

The pseudoscience of physiognomy, in which personality is inferred from physical characteristics, also held great appeal to the

Greeks. Here was another way of getting the world to measure up. Aside from its views on foxes, the *Physiognomonica* (a treatise now fittingly attributed to an author known as the "pseudo-Aristotle") relates how those with very fiery hair (*agan purroi*) are rascals (*panourgi*), while very white skin (*agan leukoi*) was a sign of cowardice (is this a third stereotype, of the redheaded man as wimp, coming into being?). But then nothing about red hair in the Greek world was good: in *The Clouds* Aristophanes has his Chorus grumble that the state is now in the hands of "men of base metal—foreigners and redheads," and Aristophanes created more sympathetic slave characters, such as Xanthias in *The Frogs*, written *c.* 405 BC, than most. Otherwise slaves in Greek drama were grumbling, self-pitying, oversexed (masturbation in the Greek world was a vice of slaves and foreigners), uncouth, dim-witted, clownish, lazy, dishonest, and petulant. And, on stage at least, they had red hair.

The Roman encounter with the redheaded world tended to be at the sword's point, rather than within the *theatron*.

Red hair is liminal, as has been said (this of course only contributes to its status as "other" to begin with). It's out on the edge, geographically as well as genetically, with an undisturbed gene pool giving it, as a recessive gene, the best chance of coming up. It's like blackjack: if you're playing the same cards over and over, sooner or later, you'll get a natural twenty-one. If there truly were unusually large numbers of redheads in Thrace, this might be why;

all those separate tribes unable to agree, or presumably to inter-marry, between themselves. You find redheads throughout Europe, but as a rule of thumb you find them in greater numbers the farther north you go. By far the longest-lasting of the tribal civilizations encountered by the Romans was that of the Celts, whose lands, at their greatest extent (depending on where you place the borders of the Celtic language, and where you place the borders of what was, and was not, Celtic in the first place) can be said to have reached from Ireland to Poland, as far north as the Hebrides, as far south as what would become Genoa. As the Roman legions marched and conquered and pushed the borders of the Empire ever up into the Celtic heartlands of Germania and Gaul, they encountered more and more redheads among the peoples they subjugated. Caesar's Gallic campaigns of 58–51 BC alone are estimated to have added a million slaves to the Roman Empire. And although there is a notion that blond- or red-haired Celtic slaves were prized, and although they may have been so, as novelties, there is little evidence that the indi-vidual slaves themselves were regarded as being of any extra worth at all. Rather the reverse, in fact. The philosopher Cicero (106–43 BC) writes of British slaves, "I think you would not expect any of them to be learned in literature or music." You can almost hear the sniff of disdain.

But contact with these Northern tribes did lead to the recogni-tion that people who looked so similar must be connected in some way. Here is Tacitus (AD 61–after 117) on the tribes of Britain, and he's worth quoting at length because here, finally, history has allowed the historian to catch up, and Tacitus is reflecting upon events within his own living memory. The original invasion of Britain by Julius Caesar in 55 BC led to conquest under the Emperor Claudius in AD 43 and was followed by decades more of on-and-off campaigning. Indeed Hadrian's Wall, which marked the outermost northern border of the Roman world, was begun only in AD 122. In his life of the Emperor Agricola, Tacitus writes:

> Who were the original inhabitants of Britain, whether they were indigenous or foreign, is as usual among barbarians, little known. Their physical characteristics are various, and from these conclusions may be drawn. The red hair and large limbs of the inhabitants of Caledonia point clearly to a German origin. . . . Those who are nearest to the Gauls are also like them, either from the permanent influence of original descent, or, because in countries which run out so far to meet each other, climate has produced similar physical qualities. But a general survey inclines me to believe that the Gauls established themselves in an island so near to them. Their religious belief may be traced in the strongly marked British superstition. The language differs but little; there is the same boldness in challenging danger, and, when it is near, the same timidity in shrinking from it. The Britons, however, exhibit more spirit, as being a people whom a long peace has not yet enervated. Indeed we have understood that even the Gauls were once renowned in war; but, after a while, sloth following on ease crept over them, and they lost their courage along with their freedom. This too has happened to the long-conquered tribes of Britain; the rest are still what the Gauls once were.

Not quite as long-conquered as Rome might have thought.

The counties of Suffolk, Norfolk, and Essex, in the east of England, exist under what Tacitus describes as a "sky obscured by continual rain and cloud." The shoulders of the fields are dotted with Bronze or Iron Age tumuli that predate even the Romans, the churches are mostly medieval, the towns are small, the villages are tiny. The landscape dwarfs the people. It feels old. The skies go on forever, and dwarf everything. And as every East-Anglian schoolchild knows, in Roman times these were the homelands of the Celtic tribe of the Iceni, and of their queen, Boudicca. To such children (I was one myself) Boudicca's chariots, with the terrifying scythes mounted Persian-fashion on their wheels, ready to cut down her enemies, are every bit as real and realizable as the tractor in the next field, grumbling home.

It was Rome's policy to move from subjugation to colonization to integration, and there were native Britons, especially in the south and east of the country, who accepted the conquest, the changed state of their world, recognized the emperor in Rome, and prospered and grew rich on Roman loans. Boudicca's husband, Prasutagus, was one of these. At the same time, again according to Tacitus, all the tensions one would expect in a newly conquered territory were present: there were Roman veterans living around the colony of Camulodunum, now Colchester in Essex, who had taken over native lands and home-steads and were especially hated, while even Roman slaves found in the native population people they could patronize and insult. Then Prasutagus died, the loans were called in, his lands taken away from his family, his wife was scourged, his daughters raped, and in AD 60 the Iceni rose in revolt. And thus Boudicca has come down to us, in the words not of Tacitus this time but of a later historian, Cassius Dio. He describes her as "possessed of greater intelligence than often belongs to women. . . . In stature she was very tall, in appearance most terrifying, in the glance of her eye most fierce, and her voice was harsh; a great mass of red hair fell to her hips; around her neck was a large golden necklace; and she wore a tunic of divers colors over which a thick mantle was fastened with a brooch."

Again, the evidence has to be sifted. Cassius Dio lived AD 155–235, so his account is certainly not as contemporaneous as Tacitus, and shows it, in its elaboration. What both sources agree upon is that the Iceni and their neighbors the Trinovantes joined forces, and with Boudicca at their head swept down on Camulodunum, then on the new trading post of Londinium, and then Verulamium, present day St. Albans, too, and razed them to the ground. Eighty thousand Roman citizens are thought have perished before the rebellion was put down, and in fact both Cassius Dio and Tacitus speak of the distinct possibility of the island of Britain being lost to Rome altogether.

Cassius Dio, like Xenophanes, wrote in Greek, and that one word, "red," in his description of Boudicca is also translatable as tawny, or as reddish-brown. Nor is there any evidence that scythe-bearing chariots were used by the Britons. Yet in the sculpture set up in 1902 on the corner of Westminster Bridge and the Victoria Embankment in London, there Boudicca is, in all her stern-browed Victorian glory, with knives a yard long protruding from the hub of her chariot's wheels. And, right up to Marvel Comics' *Red Sonja*, red hair has been indispensable to the image of the indomitable, ferocious, and usually voluptuous female barbarian. Because, again, what other color could Boudicca's hair be? In her case it speaks of her unvanquished determination, her patriotic courage (one reason why she was so popular with the Victorians), her non-Roman-ness. Red hair meant Celt, it meant Gaul, it meant Thracian. Sometimes, indiscriminately, it meant all three.

The Capitoline Museum in Rome contains a sculpture now known as *The Dying Galatian*, but for centuries after it was unearthed in Rome in 1623, it was known as *The Dying Gladiator*, or *The Dying*

Gaul. It's a Roman marble copy (that process of historical trans-mutation again) of a lost Hellenistic original, thought to have been cast in bronze and commissioned by Attalus I, king of Pergamon in Turkey from 241 to 197 BC. It shows a naked warrior, half-lying on his shield. Scattered about him are his sword, his belt, his war trumpet. Around his neck there is a twisted, Celtic-looking torque, and his hair is in the short, punky, lime-washed spikes that must, one imagines, be just as it would have appeared and exactly as the Greek historian Diodorus Siculus, who wrote between 60 and 30 BC, describes it.[25] The man's head is bent; he props himself up on one arm. There is a wound, a sword-thrust, bleeding, under his right breast, and the sculptor somehow managed to catch the one still, central moment, when the man knows death is upon him but has not yet tumbled beneath it. Lord Byron saw the sculpture and was moved to include it in *Childe Harold's Pilgrimage*:

> *I see before me the Gladiator lie:*
> *He leans upon his hand—his manly brow*
> *Consents to death, but conquers agony . . .*
>
> *. . . his eyes*
> *Were with his heart, and that was far away;*
> *He reck'd not of the life he lost nor prize,*

---

25 In his vast *Bibliotheca Historica* Diodorus Siculus wrote, "The Gauls are tall of body, with rippling muscles, and white of skin. Their hair is blond, and not only naturally so, but they also make it their practice by artificial means to increase the color which nature has given it, for they are always washing their hair in lime-water, and they pull it back from the forehead and back to the nape of the neck, with the result that their appearance is like that of satyrs and Pans, since the treatment of their hair makes it so heavy and coarse that it differs in no respect from a horse's mane." In fact the hair of *The Dying Galatian* was recarved, perhaps with this description as a guide, in the seventeenth century.

But where his rude hut by the Danube lay,
THERE were his young barbarians all at play,
THERE was their Dacian mother—he, their sire,
Butchered to make a Roman holiday—
All this rush'd with his blood—Shall he expire,
And unavenged?—Arise! ye Goths, and glut your ire!

But the dying warrior is not a gladiator, nor a Goth, nor a Gaul. He's a Thracian, and we are back where we began.

The commission from Attalus I for the original bronze is thought to have been in celebration of his victory over a force of marauding Thracians who had settled in Galatia, in the highlands of Anatolia, and from there created such terror that they were able to extort tribute from as far afield, apparently, as the kingdom of Syria. Livy—the final Roman chronicler in this chapter—describes them thus:

> Their tall stature, their long red hair, their huge shields, their extraordinarily long swords; still more, their songs as they enter into battle, their war-whoops and dances, and the horrible clash of arms as they shake their shields in the way their fathers did before them—all these things are intended to terrify and appal.

In fact he puts these words into Attalus's mouth, as he exhorts his troops to action against the "Gauls." By the time Livy was writing, in the first century BC, it didn't matter if you were Gaulish, Celt, or Thracian. You had red hair, you were barbarian—that was all.

In 2014 workmen digging in the center of Colchester, down through the layered past, down to the level of the two-foot-thick seam of blackened rubble that still marks where Roman Camulodunum stood and burned, unearthed a horde of gold jewelry—the treasured possession of some Roman woman, hidden under the floor of her home as Boudicca's warriors descended on the town.[26] The floor of the room in which it had been hidden was strewn with the remains of food, carbonized by the heat of the fire that had destroyed the building; and mixed in with this were fragments of human bone—a piece of jaw; a piece of shin. The reality of Boudicca's attack on Camulodunum is grotesquely at odds, as these things always are, with an image of fierce and indomitable courage, with Byron's notion of innate nobility, refusing to bow beneath the oppressor's yoke. The actions of Boudicca's troops were as savage as anything perpetrated today, with particular brutality meted out to the Roman women she took captive: dragged to ancient sacred groves outside the town and mutilated with a vindictiveness that it seems almost impossible to countenance any one woman inflicting on another (purportedly the Roman women's breasts were cut off and sewn to their mouths), and then they were killed. We don't know the fate of the woman who once lived in this building in Roman Colchester, but she would have been no more an equal being to those she hid her jewelry from than they would have been to her. And she did not live to retrieve it.

---

26 See http://www.telegraph.co.uk/history/11074055/Unearthed-a-golden-Roman-hoard-hidden-from-Boadiceas-army.html.

DIFFERENT FOR GIRLS

Thus human beings judge of one another,
superficially, casually, throwing contempt on one
another, with but little reason, and no charity.

BARONESS ORCZY, *THE SCARLET PIMPERNEL*, 1905

T he nineteenth-century anthropologist John Munro, who published his *The Story of the British Race* in 1899, makes much of a supposed difference in hair color between the Vikings and the Saxons. "The Danes were distinguished by their red hair and fiery temper," he writes, "from the more phlegmatic Anglo-Saxons with light brown or flaxen hair and blue eyes." This idea of some atavistic race-memory accounting for the suspicion in which red hair was held crops up very frequently in the nineteenth century (and down in Hampshire, was clearly still going strong in my grandmother's day), but given the mixing and mingling that had gone on between the tribes of northern Europe for all those centuries before, it seems unlikely that the average Saxon and the average Dane would have varied in appearance by much at all. The Vikings were no less flaxen or any the more redheaded than those they terrorized (and never mind Hagar the Horrible and his bristling red beard). It was the new and sudden nature of these seafarers' raids that made them so feared.

The first Viking raids upon the Christian West came in the summer of AD 793, on the monastery community of Lindisfarne, an island off the coast of the north of England, and in one form or another Viking raiders alternately terrorized or settled (or both) coastal and river communities throughout Europe for at least the next three hundred years. The Vikings settled in Iceland, hence its appearance on the Redhead Map of Europe. In Greenland, the first community was founded by Erik the Red, whose name almost certainly commemorates the color of his hair, while the story of the slow extinction of one such Viking settlement there, Herjolfsnes, provides a fascinating if appalling example of what happens to an isolated community as its inner stores of Vitamin D are steadily depleted.

By the eleventh century, the Vikings had reached Newfoundland, which they regarded as the last land before one's Viking longboat tipped into the abyss, in an expedition led by Erik the Red's son, Leif Ericson. They certainly sent emissaries to Byzantium, and there is archaeological evidence that Viking traders reached Baghdad (did they add their redhead genes to those already in the local population?). There are also intriguing legends among the Paiute people of Nevada of their tribe's encounter, long ago, with a redheaded enemy around the Lovelock Caves, which have led some, rather rashly, to assume that Vikings must have penetrated far into the North American continent as well. The Vikings of Sweden raided and traded down the Volga, no doubt meeting many a remnant Scythian or emergent Udmurt population as they did so, and perhaps giving what we know as Russia its name, from a Slavic version of theirs: "Rus," or "the men who row." Norwegian Vikings took captives from the coastal settlements of Ireland and the west of Scotland, adding these Celtic genes to their own, and no doubt left descendants in these places, too. They were only finally defeated in Ireland by the great Brian Boru, at the Battle of Clontarf in 1014. But to present some race-memory of Viking raiders as the reason for the dislike and distrust of red hair is to ignore far more active and pertinent prejudices that still exist today, just as they did in John Munro's time, and which in medieval Europe were even more virulent and deeply rooted.

Moreover in England at least, Viking raiders from Denmark reached an accommodation with the indigenous population, establishing a kingdom for themselves in the northeast that stretched from York almost down to London—a third of the country, contained by the Danelaw, and subject to the treaty signed with Alfred the Great in about 880. They seeded their own language, place names, and culture, along no doubt with a few supplementary redheads, throughout northern Britain (my own sandy-haired, blue-eyed grandfather was a Cougill, from Lancashire), a region where William the Conqueror would encounter seething resistance for years after his conquest of England in 1066, and which he would viciously repress. And, of course, in the tenth century they would colonize Normandy. William's own great-great-great-grandfather was Rollo, or Hrólfr (c. 846–c. 931), who was either Danish or Norwegian, depending upon which source you believe.

The Bayeux Tapestry, marking William's victory over the English at the Battle of Hastings, and which could be contemporary evidence for the color of his hair, is not much help here, identifying various figures as *WILLEM* or *WILLELMO DUCI*, who are both sandy and brown-haired. But William's third son and the heir to the English throne, the blessedly short-lived William Rufus, seems to have owed his name either to his hair or to the florid complexion long regarded as accompanying a boorish temper.[27] The associations created in the Classical world between red hair and a suspect character were clearly still going strong, as were those linking red hair, especially where it came with a ruddy, weather-beaten skin, with low birth or as a signifier of vulgarity—if not still slave, then certainly serf. The historian Ruth Mellinkoff cites an example from the ninth-century *Life of Charlemagne* in the cautionary tale of an ill-mannered peasant who refuses to uncover his head in church. When his cap is finally dragged from his ears and his head is exposed, the priest thunders from the pulpit in a final inculpatory denunciation, "Lo and behold, all ye people, *the boor is redheaded.*" The same association would be exploited by Chaucer for the character of Robin, the drunken loudmouth miller in the *Canterbury Tales* (1380s–90s), with his spade-shaped beard as red as "anye sowe or fox," and the wart on his nose, sprouting bristles as red as those of a sow's ears. Even in Scandinavia, golden-blond Odin, Mellinkoff suggests, was the god of the nobility; redheaded Thor with his hammer was the god of the laboring man. You could rule a kingdom, in fact, and your red hair would still be used to denigrate you.

Roger I, who had taken Sicily from the Muslims by 1091, was but one of many generations of Norman mercenaries fighting in Italy, and his descendants would rule Sicily for the next century (hence the

---

27 William Rufus, "hated by almost all his people and abhorrent to God," according to the *Anglo-Saxon Chronicle*, died on August 2, 1100, while hunting in the New Forest in Hampshire. A stag crossed his path, he shouted to his companion to shoot—and the man did. William's body was left where it fell; a peasant found it and brought it to Winchester in his cart.

blue-eyed, red-haired, gloriously freckled Sicilian brother and sister I encountered in a language school in Cambridge). Frederick II, Roger I's great-grandson, extended that rule across Italy and Germany and even as far as Jerusalem. Frederick spoke six languages, launched two crusades, and was elected Holy Roman Emperor, a role in which he clashed with the papacy so frequently that he was excommunicated four times over and once denounced as the Antichrist. He kept a menagerie and, according to his enemies, a harem. He was an empiricist in matters of science, investigating the process of human digestion by practicing what was basically vivisection on his dinner guests; and a skeptic in matters of religion; and was proclaimed by his contemporaries as the wonder of the world—"stupor mundi," which somehow sounds even more impressive. Then you read the Syrian chronicler Sibt-ibn al-Jawzi's description of him: "The Emperor was covered with red hair and was bald and myopic. Had he been a slave he would not have fetched 200 dirhams at market," which seems a little harsh, however true. Frederick spoke Arabic and was one of the few European rulers of the time to approach the Muslim kingdoms of the Middle East with interest and respect. He also stood up for Sicily's community of Jews.

There had been Jewish communities living around the Mediterranean as part of the Roman Empire, and the first Jewish communities in France and Germany may date back to the same early period— indeed, possibly even predate it. There is circumstantial evidence that Jewish settlers in Spain were trading with the Phoenicians, which would have been centuries before the creation of that small, insignificant Roman settlement on a gravelly ford of the Tiber in 753 BC. By the seventh century AD there were Jewish settlements spread as far afield as China, very likely of travelers and traders who had followed the Silk Road. In the eighth century the kingdom of the Khazars represented for a century or so a Jewish power base between the Caspian and Black Seas and was a major crossroads between Russia, Byzantium, and traders from the Middle East. The first documented communities of Jews in England are perhaps the only people in all its flotsam-and-jetsam population who might genuinely claim to have come over with the Conqueror, or at least to have been brought over, from Rouen, under William's watchful eye, in 1070. William had a new kingdom to subdue, and a mighty program of castle-, or rather fortress-, building to fund. The man who could come up with the concept of the Domesday Book could put a price on anything, but William needed tribute rendered in and an economy based upon coin, not Saxon barter. Bring on the first money men of Europe, the Jews enlisted by William I, to kick-start the transformation.

In 1079 another group, five French Norman Jews (and plucky souls they must have been, too), crossed the Irish Sea to set foot on what was then the extreme western edge of the known world, in a trade delegation to the king of Munster, Toirdhealbhach Ua Briain, or so the *Annals of Inisfallen* tell us. Toirdhealbhach was seventy by this date, and had spent the previous thirty years fighting, exiling, or murdering his rivals. By 1079 he was effective ruler of half Ireland. His Jewish visitors clearly knew who the Big Fellow was, but Toirdhealbhach seems to have been less certain of them. They came with gifts, but the annals record only that these visitors "were sent back again over the sea." The event does, however, open up the pleasing possibility of this grandson of Brian Boru, this aging Celtic khan, redheaded as we may surely imagine him, greeting visitors who might have been as red-haired and red-bearded as he. Were there Jewish redheads? There both were and are.

Red hair, as has been said, survives best out of the great ebb and flow of a changing population. Those circumstances are also found in communities that are endogamous—that is, that marry within their own specific ethnic group, something the Jewish population has done for centuries. There are many, many redheaded Jews, and as these communities moved into western Europe, it was a mighty cultural mischance that they brought with them a characteristic already aligned in European culture with bad character at best and barbarity at worst. One that was already perceived as apart from the norm, was already picked out and commented upon; that was, in other words, already "racialized." Or as Eleanor Anderson describes it in her thesis on discrimination against redheads today, and in the language of contemporary psychology, with the Jews of medieval Europe we see all the hallmarks of a situation where "an individual who might have been received easily in ordinary social intercourse possesses a trait that can obtrude itself upon attention and turn those who he meets away from him, breaking the claim that his other attributes have on us."[28]

---

28 Eleanor Anderson, "There Are Some Things in Life You Can't Choose . . . : An Investigation into Discrimination Against People with Red Hair," *Sociology Working Papers* 28 (2002). Anderson's thesis does a superb job of exploring prejudice against red hair as part of the overall landscape of discrimination and cultural stereotyping.

We in our communities today know that stereotyping or stigmatizing individuals on the grounds of their difference is as destructive of those who think that way as it is of those victimized by their thinking. We may not all always understand this, but we all know it, and in the First World, at least, many if not most modes of living and of belief that would have reduced our ancestors to a blood-flecked mob do so no longer. But in medieval Europe, Jews were Christ-killers and the abductors of Christian children. They were known as usurers, moneylenders, and the financiers of the state (which very often turned on them as well). And some of them had red hair. And since Judas was a Jew, in a noticeable number of examples in European art, particularly in Germany, he is also depicted as a redhead. Even the redhead's freckles were not spared. In medieval Germany, one term for freckles was "Judasdreck."[29] As the scholar Paul Franklin Baum put it in 1922, "There can be little doubt that this tradition is simply the application of the old belief—much older than Judas Iscariot—that red-haired men are treacherous and dangerous, to the arch-traitor, some time during the early Middle Ages."[30]

You read with sinking heart the first edicts forcing Jews to identify themselves by wearing specific badges; those moving them into ghettoes; the first instances of persecution, the first massacres, the first expulsions. Jewish communities were expelled from France in 1182; recalled in 1198 (the royal coffers were running low); expelled again in 1306 and their property confiscated for exactly the

---

29 Which basically translates as "Judas-shit." See Ruth Mellinkoff, *op. cit.*, p. 168.

30 Paul Franklin Baum, "Judas's Red Hair," *Journal of English and German Philology* 21, no. 3 (July 1922): 520–29.

same motive. There were no Jewish communities in England after 1290; and in 1492, in Sicily, which was under the control of Ferdinand and Isabella of Spain, those same Jewish communities that had been defended from the zeal of the Crusaders by Frederick II were driven out of the island altogether. In that year the redheaded Genoese sea captain Christopher Columbus followed Leif Ericson across the Atlantic in an expedition funded by Ferdinand and Isabella, and the world changed once again. Those monarchs, newly in control of a kingdom from which any religion other than Catholicism was to be extirpated, also, notoriously, put into motion the diktats of the Spanish Inquisition.

Historians have argued back and forth over the number and nature of the victims of the Inquisition in Spain and its territories in the New World, but however many and whoever died because of it, their suffering was inflicted by a bureaucracy of terror that used symbols of "other" as a means of identifying its victims. Trials were secret; punishment was public and vindictive enough to extend even to the burning of corpses. The Inquisition was not directed solely against those of faiths other than Catholicism, either; it went after homosexuals, those in possession of prohibited texts, and those suspected of witchcraft. In the nineteenth century it was hunting down those "suspected" of freemasonry, as well, which rather starts to sound as if there was no one else left. Right from the start it was suspected that those persecuted were being chosen because they were wealthy, and the state could confiscate their property. The processes of the Inquisition were as absurd as they were appalling, as such systems always are. But if red hair meant you were Jewish (it might equally have meant you were Protestant, or merely un-Spanish in some way), red hair might also mean you were a backsliding *converso*, especially given all those other qualities ascribed to redheads concerning their treacherous and untrustworthy nature. You end up going around in a circle here, where the prejudice justifies the racism and the racism strengthens the prejudice. Attitudes toward red hair in Spain only reflected the fears and prejudices of Catholic Europe as a whole. If you want to look for reasons for the continuing and increasing antipathy toward redheads, in particular redheaded men, in medieval Europe, look no further than its anti-Semitism. And if you want to place the point at which attitudes toward red hair in men and women begin so radically to diverge, likewise.

To understand what a people thought or believed, start by looking at what they were looking at. In this case, two panel paintings, one German, one French, created less than ten years apart, demonstrate the polar opposites of attitudes toward red hair in the medieval world. The German work shows Christ on the Mount of Olives, or *The Agony in the Garden*, as the scene is perhaps better known. It dates from about 1444–45 and is now in the Bayerisches National-museum in Munich (Fig. 7). For many years—centuries, in fact—it was unattributed to any particular artist; another of those anonymous works that in their lost history speak with quiet eloquence of the unknown, uncountable faithful who knelt in front of them. Its creator was identified only as late as the 1980s as Gabriel Angler

the Elder, a Munich artist whose dates are thought to have been
c. 1405–62.[31] Previously his works were simply grouped together
under the title of the most famous of them, and he was known as the
"Master of the Tegernsee Altar."

In his painting, we are in a garden of sorts, although parts of
it seem very rocky and untended. It's bounded by a wicker fence,
which might bring to mind images of that first garden, Eden, or
of unicorns and virgins, or indeed of any place private, secluded,
and reserved for contemplation or for prayer. And it's night. There
are four figures in the garden, three of them huddled in sleep, and
although one should be aware that the background was repainted
in the late seventeenth century, the re-painter, in his creation of a
sky literally dark with foreboding, did a good and sympathetic job.
This seems to be one of those nights where heaven has come very
near to earth—the sky is framed by delicate Gothic tracery, and in
this realm of the divine, top left, there hovers an angel bearing a
scroll. You can almost hear the stillness—the nighttime insects, the
lazy flapping of the scroll as it unfurls from the angel's hand, the
whispering of the one man awake, kneeling at the rocky outcrop as
at a prie-dieu.

Enter, stage right, as from some other world beyond the painting
and through the flimsiest-looking portal, a half dozen soldiers in
full fifteenth-century armor; helmeted, with swords at hip, and spear
and halberd glinting above them—a pre-Reformation SWAT team.
They are being led through the fragile doorway into the garden by a

---

31 V. Liedke, "Die Münchner Tafelmalerei und Schnitzkunst der Spätgotik, II: Vom Pestjahr 1430 bis zum Tod Ulrich Neunhausers 1472", *Ars Bavarica*, xxix/xxx (1982): 1–34.

man in a long, Biblical, yellowish robe, as if he has summoned them from the artist's time into his. One of his hands is raised, a finger pointing in admonition—*shush!*—but of course also, ironically, up to the divine; while with the other he supports the weight of the heavy purse hung around his neck. This is Judas. And his hair, and his beard, and even the skin of his cheeks, are red.

Redheaded men in medieval Europe were at just as much of a disadvantage as they would have been in classical Greece or Rome (and it should be remembered, throughout this period, men's hair was much more publicly visible than women's). Those who wished to justify their prejudice might point to the Old Testament example of boorish Esau, born "ruddy, all over like a hairy mantle," and dim-witted enough to lose his birthright to his younger, more intellectual brother, Jacob. Cross such men (trick them out of their rights as firstborn over a bowl of lentil soup, for example, as Jacob did), and the dark flush of their ruddy visages betokened violence, loss of temper, retribution that would be both swift and unthinking.[32] Even the Biblical King David might be so judged: "And when Samuel saw that David was ruddy he was smitten with fear, thinking he might also be a murderer." And if King David's temper might be suspected on the basis of his coloring, how much more likely, and easier, was it to project one's fears onto one's foreign, heretical, idolatrous, synagogue-going neighbor, with his strange dietary observances and his red hair?

It's ridiculous to assume that in reality red hair was any more

32 Ruth Mellinkoff *(op. cit.)* points out that the tiny figure of Cain, a marginal illumination in the Mosan Park Bible of *c.* 1148, also has red hair. BL Add MS 14788, folio 6, verso.

common then than it is today. It was and is still a minority characteristic, among Jews just as it is among the Irish and the Scots, and
as it no doubt was with the Thracians, too. But once again it comes
to stand for an entire ethnicity of other: once viewed, apparently
blinding those who saw it to any other characteristic.[33] In medieval
art red hair in men, in particular in combination with a ruddy skin,
is like the black hat on the baddie in a Western. It's visual shorthand
for a brutal character of a particular unthinking sort—animalistic,
unintellectual, unreachable by reason, and all the more frightening
for that. It's a means of signifying not simply villein, or serf, but villain, too.[34] It's the color of the hair of Christ's (Jewish) tormentors,
with their whips and goads and what in so many medieval paintings
is the artist's invention of ridiculously, sadistically, overelaborated
means of pressing the crown of thorns down on Christ's forehead
(why not simply have his torturers wear gloves?). It's noticeable that
it's in this combination with ruddy, rough, or weather-beaten skin
that red hair appears in some of the most nastily caricatured depictions of Judas. And nakedly exposed, time and again, as Mellinkoff
points out, it is the coloring of the impenitent thief at Christ's crucifixion (Fig. 8). Distorted, damned and writhing in agony, possessed
by devils, abandoned by God and man (and the Catholic church), this
is the uncouth barbarian again. The coloring of the impenitent thief,

---

33 Or as Bishop Jocelin puts it in William Golding's *The Spire* (1964), on seeing Goody Pangall's red
hair, "It was as if the red hair, sprung so unexpectedly from the decent covering of the wimple, had
wounded all the time before, or erased it."

34 And as such, well into the nineteenth century, it stigmatizes Quasimodo in Victor Hugo's
The Hunchback of Notre Dame (1831): "a huge head bristling with red hair . . . that little left eye obstructed
with a red, bushy, bristling eyebrow." Sure enough, Quasimodo's looks are taken as proof that he is
"as wicked as he is ugly." He even remains a redhead in the Disney version of 1996.

by contrast to the pale, serene, penitent thief, heading for Paradise, is used both to evoke his freakishness and his inhumanity, the fact that his fate doesn't matter, and to demonstrate, with the artist as our proxy, as it were, society's and our Christian revenge.

And oh, how that prejudice linking Judas and Jewishness and red hair persisted. In *As You Like It*, written in 1599, a century and a half after the completion of Angler's painting, Shakespeare has Celia describe Orlando's hair, defensively, as "something browner than Judas's." In the 1690s, in the midst of a bitter falling-out with his publisher Jacob Tonson (a businessman canny enough to have put the first copyright on Shakespeare's plays), the poet John Dryden anathematized Tonson as having "two left legs and Judas-colored hair." The association between red hair and Jewishness was so strong that Shylock was still being played in a red wig until Edmund Kean performed the part at Drury Lane in 1814.[35] Even Charles Dickens, who could make as impassioned a case for the dispossessed and the minoritized as any writer before or since, in his 1838 novel *Oliver Twist* gave the world the character of Fagin, "a very old, shriveled Jew, whose villainous-looking and repulsive face was obscured by a quantity of matted red hair." Fagin's character very likely contributed at least as much to the negative associations of red hair as the prejudices that gave us red-haired Judas in the first place.[36]

---

35 John Gross, *Shylock: A Legend and Its Legacy* (New York: Simon & Schuster, 1994).

36 In his defense, Dickens did tone down the anti-Semitism in later editions of the book, and in his public readings of the part, but by then, one might argue, the harm was done. It's also intriguing if again deeply depressing to speculate how much the belief in the Jews as the exploiters of Gentile children might have influenced the creation of Fagin's "gang" of child thieves and pickpockets, too.

And how it still persists. The notorious "Ginger Kids" episode of *South Park*, broadcast in 2005, with Cartman proclaiming that "Gingers have no souls," and can't go out in sunlight (much to the fury of Kyle, possessor of a little-seen but magnificent red "Jewfro"), was only mining a much older tradition that Judas became a vampire and roamed the world as one of the undead. It's pretty much a perfect example of how such folklore accretes, one layer around another, with our own age merely adding the most recent. In 1887 the French geographer Élisée Reclus recorded a Romanian belief that "if the deceased has red hair . . . he would come back in the form of dog, frog, flea or bedbug, and . . . enter into houses at night to suck the blood of beautiful young girls." Vampires, so the thinking goes, are bloody; red is the color of blood, therefore redness must predispose one to vampirism.[37] Indeed, until John Polidori, Lord Byron's physician (and uncle, incidentally, of the pre-Raphaelite artist Dante Gabriel Rossetti), published his genre-defining short story "The Vampyre" in 1819, rather than being the pale-skinned, dead-eyed aristo embodied by the mysterious Lord Ruthven, vampires were given away by the ruddiness of their countenance, and the way in which it suggested that they were gorged with blood.

In fact the association between red hair and vampires noted by Reclus might hark back to even older folklore, as far back as Byron's Dacians, in fact, who were another Indo-European tribe in the region west of the Black Sea, and who may have been Thracian, originally, but with a greater influence from Scythia and the

---

37 For this and further detail on the evolution in European folklore of the undead, see Paul Barber, *Vampires, Burial, and Death: Folklore and Reality* (New Haven, CT: Yale University Press, 1988).

Celts. Or they may not. It's as well to take all this with a pinch of salt, especially as one of the earliest writers to interest himself in this area was one Montague Summers (1880–1948). Summers's personal *shtick* was to present himself as an academic witch-hunter, as opposed to the insane necromancer role inhabited by his notorious contemporary Aleister Crowley. (The two were acquainted, naturally enough—Summers was very possibly a Satanist himself, and very likely a pedophile as well.) Summers traces a Greek belief that those who commit crimes in life return as vampires to Slavic mythology, and he states that people with red hair and gray eyes are regarded as vampires in Serbia.[38] There are other associations as well that help explain this particular bit of mythology. Judas's putative wanderings as a vampire, rejected by both heaven and hell, mimic those of the Wandering Jew, while the vampire's invulnerability might suggest the Old Testament narrative of the Mark of Cain. The silver bullet that supposedly downs witches, werewolves, and everything else that goes bump in the night could suggest the thirty pieces of silver paid to Judas for betraying Christ. It all goes to show that you need neither the internet nor a butterfly to create a woozle.

The *South Park* "Ginger Kids" episode was the catalyst at a school in Yorkshire (more Northern redheads) for a seriously unpleasant "Kick a Ginger Day." So if this story has a moral, it's this: don't show sophisticated satire on the stupidity of racism to the stupid. Or as Ruth Mellinkoff puts it, "Red hair is a minority feature, and this fact is sufficient to explain why it is used in the visual arts as a

---

38 Montague Summers, *The Vampire in Europe*, reprinted Aquarian Press, 1980.

negative attribute and why it is still widely treated with suspicion. Antipathy to red hair, a red beard, and ruddy skin, is as simple, and as complicated as that."

Quite.

The French panel-painting is different from the German work in every way. It shows *The Coronation of the Virgin* and was commissioned for the Carthusian monastery of Chartreuse du Val de Bénédiction, Villeneuve-lés-Avignon, where it remains to this day, and where we can assume it has been since September 29, 1454, the delivery date specified in the contract for the work, drawn up between the clergyman who paid for it, one Jean de Montagny, and the painter, Enguerrand Quarton (Fig. 9).

The survival of Enguerrand Quarton's contract for his *Coronation*, and the astonishing detail this document goes into about the commissioning of the work, is one reason why the painting is so well known today—among students and scholars of medieval art, that is. (The other and much better reason is of course the painting's peerless, nigh-on flawless, beauty.) It is unusual for any document of this period to survive. Scholars of medieval art spend years scouring archives and parish records for a single mention of the artist they are researching, and the discovery of even the most abbreviated reference will generate years' more speculation and analysis. In this context, the commissioning document for Quarton's *Coronation* has the same kind of importance as the discovery of the El Sidrón family of Neanderthals. Among other details that the commission specifies are these: the overarching composition of the painting (there should be, it says, with unconscious poetry, "the form of Paradise"), the specific identities of the major saints to be included in that Paradise, that the Mass of St. Gregory was to be shown in the cutaway church lower left, and that in the representation of the Holy Trinity "there should not be any difference between the Father and the Son." It specifies the payment stages and how long the artist would have to complete the work. It even records where the document was drawn up and signed: in the spice shop of one Jean de Bria. It describes what the Virgin Mary should be shown wearing ("white damask"), and that the Holy Trinity should be surrounded by cherubim and seraphim, as indeed in the finished work they are. Yet in all these specifications it leaves the details of the appearance of the Virgin, the central figure of the painting, to the artist, "as it will seem best to Master Enguerrand."

In fact no fewer than six separate contracts for Enguerrand Quarton have been found, although only two of the works they describe have survived. We know or can extrapolate from the documents that he must have been born around 1420, that he was in Aix-en-Provence in 1444, Arles in 1446, and rented a house in Avignon in 1447. He seems (unusually, for an artist of this or perhaps any period) to have delivered the works commissioned from him without default. As far as we can judge, he moves seamlessly from locale to locale and patron to patron, from one work to a larger work, and from one fee to a greater fee. His first commission in Avignon was for a *Madonna of Mercy* for the Celestine convent there (the painting is now in the Musée Condé in Chantilly); the *Coronation* followed five or six years later.[39]

Everything about this painting is special. There is its glorious condition, to begin with. There is Quarton's palette—the fiery reds and deep cool blues for heaven, the pastels for our world beneath. There are the saints and priests and bishops, the holy martyrs, each one in this cast of hundreds differentiated from the next, each face as real and fully realized as any you might encounter in Avignon today. There are God the Son and God the Father, an almost identical pair of bearded brothers, clothed in red silk lined with green. And between them there is the Virgin, shown as seemed best to the artist. And how did she seem best to him? Not in white damask, for sure. Despite the stipulation in the contract her robe is cloth-of-gold, shot with a bold pattern of stylized blooms. And her hair is red.

There she kneels, on a kind of ermine of fluffy cloud and blue cherubs with her aristocratic, long-fingered hands crossed over her breast. Her head is tilted beneath the fluttering wings of the Holy Spirit and the weight of the crown being placed on her head by God the Father to one side and her son to the other. Her eyes are half lowered and her bright red hair is falling over her shoulders. In fact such a redhead is she that the color of her hair and the ivory of her skin can hold their own against even the background of burning seraphim. Nor as a redhead is she alone in Quarton's works. The Virgin of his *Madonna of Mercy* also has red hair visible under the cloak with which she shelters the faithful. She also has the same oval face, point

---

39 The last occurrence of his name is in 1466. The plague came to Provence in that year; Quarton, it is thought, may have been one of its victims.

to her chin, long, straight nose, fine eyebrows, and narrow, slightly Asian-looking eyes. Was she simply Quarton's ideal? Looking at the face of the Virgin in the Coronation, some might think they can see personality, challenge even, in the gaze with which she meets the viewer from under those half-lowered lids, a vivid comprehension of her role. Might that hair have been encountered for real in the streets of Avignon, six centuries ago?

Artists are among the most fervent admirers of red hair, and of the pale skin that so often goes with it. Hans Memling's triptych of *The Last Judgement* of *c.* 1467–73 has two, if not more, proudly nude female redheads, having been greeted by St. Peter, in the crowd dignifiedly walking up the crystal stairs of the painting's left wing to Paradise (Fig. 10).[40] They have their backs to us, and their rippling red hair reaches to their hips. The Archangel Michael in the same work, dividing the saved from the damned, has red hair, too. Red hair on women, on angels, is a thing of beauty, so these medieval artists seem to be telling us, except of course (they add) when it's not. So in these medieval paintings we have on the one hand the Queen of Heaven, in all her beauty and divinity, depicted as a redhead, and on the other hand we have Judas Iscariot, also portrayed as a redhead. Clearly in the one case red is good, and in

---

40 This work has perhaps one of the most adventurous life stories of any medieval painting. Commissioned by the Medicis' banker in Bruges, Angeleo Tani, it was being shipped to Italy by his successor, Tommaso Portinari (who may be the little figure in the left-hand bowl of St. Michael's dreadful scales), when the vessel carrying it had the bad fortune to run into the hands of a Polish privateer, Paul Beneke. Clearly a man of taste, Beneke presented the painting to the Church of St. Mary in Gdansk, and there, despite lawsuits demanding its return, it remained until looted by Napoleon in 1807. After Napoleon's fall, it found itself in 1817 in Berlin, from where St. Mary's Church was able after some effort to reclaim it; after World War II it turned up in Leningrad, and was finally returned to Gdańsk in 1956. It is now in the National Museum.

the other it is about as bad as bad can be, and the question begging to be answered is *why? Why* is red hair so gendered? Why is it so different for girls?

Here is another painting: this time in the Musée Unterlinden in Colmar, and that enjambment of French and German in the museum's name should hint at a good deal of Colmar's history. Founded in the ninth century, the city nudges up against the Franco-German border, which has swung back and forth over Colmar like a jump rope. It was a part of the Holy Roman Empire until 1673, when it was conquered by Louis XIV. It came back under German rule in 1871, as a result of the Franco-Prussian War, returned to France in 1919, was annexed by Nazi Germany in 1940, and returned to France again in 1945. With so much war being waged around it for so long, you might think very little of old Colmar would survive, but you'd be wrong. The city is as pretty as a film set, its elaborate timber-framed architecture, with witch's hat roofs and bold carved sandstone doorways, suggesting quite how much money might be made out of being a border trading post in the fifteenth and sixteenth centuries.[41] The wines are delicious, the cuisine heart-stoppingly good, the people pragmatically bilingual. In 1834 it was the birthplace of Frédéric Auguste Bartholdi, designer of the Statue

---

41 The Maison Pfister in Colmar was in fact in a movie—it is the inspiration for *Howl's Moving Castle* in the eponymous animation from Studio Ghibli (2004).

of Liberty, and in 1450 or thereabouts, the birthplace of Martin Schongauer, whom one may speak of as the foremost German artist of his generation, and whose workshop (every successful artist of the period created a workshop of assistants and apprentices) in *c.* 1480 created the panel painting shown in Fig. 11. It too blurs boundaries.

Once again, we are in a garden, but a very different one from that of Gabriel Angler's *Betrayal.* The wattle fence is solid, the gate is of knotty two-by-four, painted with such solid realism you could build a copy of it today. Like the lych-gate to a church, it has a little roof. The sky is neither dark with night nor overpaint, but gold—this is a holy space—and the ground is green with grass as lush as fur. It might be a latter-day Garden of Eden. There is a rose bush, blooming, what may be a tree peony. There are birds in the branches of the trees. One looks like a wagtail and the other a chaffinch, in which case, because no detail in paintings of this period is accidental, the wagtail may symbolize earthly love and beauty (the bird was linked in Classical mythology to Aphrodite), while the chaffinch was a symbol of celibacy. They create a counterpoint to the figures in the foreground, the Adam and Eve of this place: a woman kneeling on the grass, her hair in thick coppery ringlets, and a man, looking back at her even as he walks away. The man is Christ, carrying the banner of the Church and dressed in a triumphant scarlet toga, arranged to display the sacrificial wound in his side (the same wound, incidentally, as the *Dying Galatian*—how these things repeat themselves); the woman is Mary Magdalene. This is the *Noli me tangere,* the moment after the Magdalene's recognition of the risen Christ outside the tomb. For my money it is one of the most psychologically complex moments in all Western art, and the figure of the redheaded

Magdalene, as Western art and literature has created her, one of the most multilayered and compelling—and, I would argue, the single most important reason why Western attitudes toward redheaded men and redheaded women diverge so thoroughly. Accidents of history have given redheaded men the stereotypes of barbarian warriors, clowns, milksops, arch-traitors, or violent God-damned brutes. It has given redheaded women Mary Magdalene.

The Magdalene as she has come down to us today is a conflation of at least four figures from the Biblical story. There was Mary Magdalene herself, present at both the Crucifixion and the Resurrection, and the first to whom the risen Christ revealed himself. There was Mary of Bethany, sister of Martha and Lazarus, who according to the Gospel of St. John had anointed Christ's feet with perfume and wiped them with her hair. And there is the unnamed woman who in Luke's gospel approaches Christ in the house of the Pharisee and does the same, and who is described only as "a sinner." There was another Mary, whom Christ cured of demonic possession in the Gospel of St. Mark. Finally, Mary Magdalene's medieval biography, in its best-known telling, by the thirteenth-century churchman Jacobus de Voragine, seems to owe something to the story of Mary of Egypt, another prostitute-saint, who wandered the desert clothed only in her hair.

The problem with the Bible, certainly for the early fathers of the Christian church, who saw truth as having to be one narrative or as being nothing at all, is that so often the sources they were teaching and preaching from were various and contradictory. Nor was this a problem confined to Christianity. The Talmudic *Alphabet of Ben Sira*, of the eighth to tenth century, conjures into being a first wife for Adam, pre-Eve, to explain the fact that in the Book of Genesis a "wife" is mentioned twice: first at Genesis 1:27:

> So God created man in his own image, in the image of God created he him; male and female created he them.

and then at 2:22:

> and the rib, which the LORD God had taken from the man, made he a woman, and brought her unto the man.

Out of this minor inconsistency, and a good deal of padding from earlier Jewish and even Babylonian mythology, was created the legend of Lilith, the woman who, seeing herself as her husband's equal (unlike docile Eve, borne of his rib), refuses to "lie beneath," quarrels with Adam, and strikes out into the Babylonian wilderness on her own. Disputatious and disobedient therefore, and yes, to this day, very often depicted with red hair.

Such painstaking elaboration to explain what was basically even less than a typo might strike us as hardly worthy of the effort, but to the first Catholic popes of the post-Classical world, pressed by the Goths on one side and the power of Byzantium on the other, yes, it was. In 591 Pope Gregory the Great (he of the Gregorian chant) decreed in his Homily 33 that the unnamed sinner of the Gospel of St. Luke, Mary of Bethany, the woman cured of seven devils in the Gospel of St. Mark, and the Mary Magdalene of Christ's earliest followers were one and the same.

> She whom Luke calls the sinful woman, whom John calls Mary, we believe to be the Mary from whom seven devils were ejected according to Mark . . .

. . . It is clear, brothers, that the woman previously used the unguent to perfume her flesh in forbidden acts. What she therefore displayed more scandalously, she was now offering to God in a more praiseworthy manner. She had coveted with earthly eyes, but now through penitence these are consumed with tears. She displayed her hair to set off her face, but now her hair dries her tears. She had spoken proud things with her mouth, but in kissing the Lord's feet, she now planted her mouth on the Redeemer's feet. For every delight, therefore, she had had in herself, she now immolated herself. She turned the mass of her crimes to virtues, in order to serve God entirely in penance.

And what a winning combination this sinful woman, this "peccable female," was to prove to be.

According to Jacobus de Voragine in the *Golden Legend,* his bestselling compilation of the lives of the saints, Mary Magdalene was the daughter of wealthy parents, and "shone in beauty," as de Voragine puts it (or rather as William Caxton translated him, in 1483). Her sins were her pleasure in her looks and her riches, and her giving up of her body to "delight," rather than to straightforward prostitution. Female beauty, female wealth, and, above all, female delight were deeply suspect to the medieval Christian church (there are times when it feels they are hardly less so today), and sure enough, they are Mary Magdalene's downfall. Inspired by the Holy Ghost, she is filled with remorse for her sins, renounces her wealth, buys a pot of the costliest ointment (it's the object looking like a miniature fire hydrant by her knee in the Colmar panel) and, penitent and weeping, approaches Christ in the house of the Pharisee. There follows the well-known story of the tears, the feet, the ointment, the hair. After the Resurrection, in a crackdown by the authorities, Mary Magdalene and companions, including St. Maximin of Aix, are cast into a rudderless boat and set adrift, washing ashore at Marseille, where she begins to preach. There is the conversion of the local prince, there are miracles and visions, and once the prince has destroyed his heathen temples, Mary Magdalene retires to "a right sharp desert," as Caxton feelingly describes it, where she spends the next thirty years in solitary contemplation, sustained only by her faith and by a daily angelic "assumption," where choirs of angels sing to her and feed her. She dies after receiving a last communion from St. Maximin, is herself anointed "with divers precious ointments," and buried. By the tenth century her relics were to be found many hundreds of miles inland, at Vézelay in Burgundy.

But holy relics were big business in the Middle Ages, and in 1279 Charles II, Prince of Salerno, announced that he had been told in a dream that the true relics of Mary Magdalene still lay within the chapel of St. Maximin, near Aix-en-Provence. Charles was involved for most of his life in the tripartite power plays between the kingdoms of Aragon, Naples, and Sicily for control of the latter (a centuries-long dispute that can be traced back to the death of Frederick II), but he was also Count of Anjou, whose territories included Aix, and the details of his dream, were they to be proved true, would give him a valuable degree of both divine and papal endorsement. So a tomb in the crypt of St. Maximin was opened, the relics were indeed discovered (unperished and exquisitely perfumed), and their approval by the pope, Boniface VIII, was sought and granted. Charles founded a massive pilgrim's basilica on the site, the saint's remains were enclosed in a gold reliquary, complete with rippling beaten gold hair, and the cult of the saint began to spread, first via Naples into Italy and then throughout Europe.

So much for the history. Let's talk about the woman.

How does one begin to unpack the symbolism around Mary Magdalene, to analyze what it was that made her so significant, so ubiq-
uitous and so beloved a figure in the medieval mind? Where do
you start? There is her narrative to begin with—the wealth, the
privilege, come to naught, rejected for something greater.[42] This is a
storytelling arc to which something very deep and very human in us
still responds today—that combination of schadenfreude evolving
into empathy and then even into admiration drives most reality TV,
for example. There is her role, even her posture, as the tearful penitent, the seeker of reconciliation, the rebellious female brought to
heel, contrite and come to make up, as if to a lover—on her knees at
Christ's feet, wiping them with her hair, or in its counterpart, kneeling in the garden, reaching up toward him—and the fact that this
scene takes place in a garden makes her also, of course, the antidote
to the first peccable female, Eve. Christ touches her on her forehead,
which might be taken as an antidote to the Mark of Cain, as well.
Then there is her own bittersweet agony of *"Noli me tangere,"* the
role reversal of look-but-don't-touch, the way in which, although it
is the Magdalene kneeling, thus apparently in the subservient posture, her reaching out toward Christ might be seen as placing him

---

42 St. Francis of Assisi, who is sometimes cited as the male version of Mary Magdalene, shares the same
kind of narrative arc to his life, and is another saint with now more than pan-European appeal.

in the role normally reserved in the medieval world for the unattainable female object of adoration or desire (you see what I mean about the Magdalene's compelling nature?). And then there are her tears. The Magdalene weeps, she feels, she has human emotions and artists depict her as giving them vent—gathering her tresses around her in Titian's *Penitent Magdalene* of 1533, now in the Palazzo Pitti in Florence, brimming eyes turned up to heaven. Or, a hundred years earlier, in Jan van Eyck's *Crucifixion* (Fig. 12, *c.* 1428, Metropolitan Museum of Art). Here the Virgin slumps bottom left, face hidden, literally shrouded in grief. Mary Magdalene is beside her with her back to the viewer (and yes, again, the fall of rippling red hair) raising her arms as if opening her body to the full agony of the scene, knitting her hands together, imploring heaven to intercede.

And then there's her sensuality and human passion.

Red hair in your female mate might be a relative guarantee of successful childbirth and healthy heirs, but there is surely more at play here than that. Red is the color of blood. One of the most ancient slurs thrown at redheads is that they are the product of sex during menstruation, in itself one of the oldest of sexual taboos. And if red is the color of blood, it is thus also the color of passion—whether of rage or of erotic arousal. And it is the color of fire. "Do not," thunders St. Jerome, in AD 403 in a letter to one Laeta, concerning her daughter Paula, "do not dye her hair red, and thereby presage for her the fires of hell," thus neatly linking all three. There seems to be no specific historic connection between red hair and prostitution, but there is a connection with the color and the profession, going back to the Old Testament, the Book of Joshua, and Rahab, the whore with a heart of gold, another peccable female who redeemed herself by hiding Joshua's spies in the roof thatch of her house in Jericho (more hair symbolism, one wonders?), and when Joshua's troops sacked the city, they hung a red cord out of her window to identify her house, and thus she and her family were spared. It's suggested that Rahab's cord may be the derivation of the red-light district, as if the color and its other associations were not already sufficient explanation. Red stands out. If as a streetwalker you wish to be noticed, it's a choice, and for many centuries it was also one of the hair colors easily achievable through natural dyestuffs, in this case henna. And it is the Magdalene's color. Rogier van der Weyden, painting Mary Magdalene in the guise of a nun in *c.* 1445, in the left-hand panel of his *Crucifixion*, now in the Kunsthistorisches Museum in Vienna, clothes her in black and hides her hair under a wimple, but nonetheless gives her a flash of red underskirt. You might speculate that artists also used the association to contrast the Magdalene with the blue robes of the Virgin. Equally it's been argued that the color red in the Middle Ages symbolized *caritas*, or the love of God (this at the same time, mark you, as it was symbolizing the complete opposite for the figure of Judas). Then there are very definite associations between long or loosened hair and sex— medieval virgins wore their hair loose, wives (whether Gentile or Jewish) wore it bundled away—indeed, one means of shaming a suspected adulteress in Jewish society was by exposing and unloosing her hair. We all know the cliché of the dowdy secretary, relieved of her glasses and with her hair out of its chignon, before whom her (male) boss stands amazed, exclaiming "My God, Miss Peabody, you're beautiful!" Is this where it began?[43] It is as if by unloosing the hair, one was setting free the sexual nature, too, and prostitutes in this period did indeed wear their hair loose on the street, as no respectable, sexually mature woman would have done. (And there were a lot of them—in the sixteenth century the diarist Marin Sanudo estimated that in Venice there were 11,000 prostitutes working in a city with a total population of some 100,000 souls.) Prostitutes were tolerated in cities such as Venice as representing a lesser vice than that of sodomy, in an age when economic circumstance forced men to marry late, and sex outside marriage all too often resulted in children born outside marriage, too. This was also the period when syphilis first appeared in Europe, very probably brought back from the New World by Christopher Columbus's crewmen, and it has also been suggested that the enormous popularity of Mary Magdalene as a saint mirrors the rise in the disease and in the number of prostitutes working the streets and at risk from it.

Yet connect the streetwalker's attribute of long loosened hair with Mary Magdalene, make it red, and its meaning is turned upside down. Instantly it becomes not only how we recognize her but socially acceptable and even desirable. Piero di Cosimo's *Magdalene* of *c.* 1500 would simply be a portrait of a woman reading at her window (here's another attribute of Mary Magdalene that I like, the presentation of her as a thinking creature, a woman with

---

43 There is a very similar meme in Indian culture, where the unloosing of the woman's knot of hair presages the unloosening of the sari, too.

a contemplative, hungry mind), were it not for the jar of ointment on the windowsill with her book, and her beautifully arranged and lushly painted long red hair, twined here with pearls. If the Magdalene is the subject, even her nudity is acceptable; and far from this being frowned on by society, is welcomed, is celebrated, putting it too strongly? Titian's Magdalene is supposedly oblivious to her naked breasts and the nipples so teasingly presented through her protective gathering up of her hair, yet we, as the viewer, can hardly be imagined so to be. We are being encouraged by the artist to gaze to our satisfaction and enjoy. This is the Magdalene as *Venus Pudica*, drawing our attention to the very thing she pretends to hide.[44] And so an age-old connection—one that we will return to again— between red hair in women and sexual desirability is pushed a little further.

And just to show how far it could be pushed, here is one final example of a painting of the Magdalene. In this she's lost her jewels, her pot of precious nard, every stitch of clothing, all habitation, every connection to the world, but if she has her red hair, we recognize her, and her original audience (almost unbelievably) found her acceptable still. In 1876 Jules Joseph Lefebvre paints her alone and naked lying at the entrance to a cave (Fig. 13). Or rather, she is his excuse for painting as erotic a nude as a Salon artist in the nineteenth century

44 There is also a winningly eccentric painting of the *Elevation of Mary Magdalene* by Jan Polack (*c.* 1435/50–1519), supposedly showing her in the desert, clothed in her hair, but instead employing and drastically misunderstanding the "feathered tights" convention of costume in medieval drama to display her instead in a little furry cat-suit, with peekaboo cutouts for her breasts. The Magdalene's hands are curved about them, in a gesture that pulls the eye toward the very thing that it pretends to hide. See http://commons.wikimedia.org/wiki/File:Mary_Magdalene_01.jpg.

could expect to get away with, and still be invited to those Salons, that is. Let's be blunt: he paints a naked woman laid out across a rock. One of her legs is drawn up; if her audience (they should be so lucky) could have moved around her even by another step she would be as totally exposed as the model in Gustave Courbet's infamous *L'Origine du Monde* (a painting which also has a unique place in the history of the redhead in art). She has a delectable body: slim-legged, round hipped, high-breasted, and her pose is calculated to display it at its most alluring. Her skin is pale, pale, pale; and Lefebvre has so arranged her that she doesn't even have a face to trouble her audience with. She's lost even that—one arm is raised to hide almost all her features but her forehead. So she's not even a whole nude; she's simply a body, save for two things: the stem of thorns that trails across her calf, and the immense spill, down to her waist, of her shimmering copper-colored hair.[45]

In the run of history redheaded women may have reason to be more grateful to Mary Magdalene than not, but we're a long way from the Queen of Heaven, that's for sure.

---

45 Lefebvre liked redheaded models, so far as one can judge. His other redheaded nudes include *Diana* (1879), *Undine* (1881), *Pandora* (1882), and the undated *Fleurs des Champs*. In 1870 he painted *La Verité*, a dark-haired model posed nude with one arm raised, who seems to have influenced Bartholdi's pose of the Statue of Liberty. Jean-Jacques Henner (1829–1905) is another French painter, more or less contemporary with Lefebvre, who favors redheaded models, but depicting them with his characteristic chiaroscuro and with rather more dignity.

Fig. 1: This portrait of a redheaded Afghan girl was taken in 2004 by REZA in the Pashtun tribal zone in Afghanistan.

Fig. 2: Uyghur girl photographed in Kashgar, in China's Xinjiang Uyghur Autonomous Region. The ethnicity of the Uyghur is Turkic, rather than Han Chinese. They see themselves as an occupied people, and identify both ethnically and culturally with the non-Chinese civilization of the Tarim mummies rather than with that of Bejing, referring to their region as "East Turkestan." Recently there has been ethnic violence in the region, bombings and knife attacks, indiscriminate as these events always are in the victims they have claimed.

Fig. 4: The red-haired divinity from coffer 32 in the ceiling of the Ostrusha tomb in central Bulgaria. Dating back to 330–310 BC, it is one of the earliest representations of a redhead in Western art.

Fig. 5: King Rhesos and a very rude awakening. This Chalkidian black-figure amphora was made in southern Italy about 540 BC. The artist, known to us as the Inscription Painter, uses a red glaze for Rhesos's hair and beard. As the story was recounted in Homer's *Iliad*, Odysseus and Diomedes infiltrated the Thracian camp outside the walls of Troy, hoping to steal their fine horses, one of which can be seen being led away to the left.

Fig. 6: This tiny terracotta figure, just 13 centimeters high, of an actor playing the part of a runaway slave, was made in Athens between 350 and 325 BC. It came to the British Museum from the collection of Eugene Piot (1812–90). Piot squandered a fortune collecting treasures from across the Classical world. Microscopic remains of red paint still adhere to the figure's terracotta hair.

Fig. 7: Gabriel Angler the Elder, *The Agony in the Garden*, 1444–45, Bayerisches Nationalmuseum, Munich. The altarpiece this panel comes from was made for Tegernsee Abbey in Bavaria, where, four centuries before, in the eleventh century, an unknown monk wrote the *Ruodlieb*, one of the earliest knightly epics, containing one of the first warnings against men with red hair: "A red beard rarely hides a good nature. . . . "

Fig. 8: Antonello da Messina, *Calvary*, 1475. The well-tended landscape, the figures of the Virgin and St. John the Evangelist, even the central figure of Christ on the Cross seem oblivious of the agonies going on around them to left and right. Note, however, the point being made by the artist in showing the impenitent thief's red hair. Royal Museum of Fine Arts, Antwerp.

Fig. 9: Enguerrand Quarton,
The Coronation of the Virgin, 1454,
Monastery of Chartreuse du Val de
Bénédiction, Villeneuve-lès-Avignon.
The Carthusian order is closed and
silent, with monks undertaking a life of
prayer and contemplation, each within
his own cell. Each cell, however, has a
garden, and once a week the community
walks in the countryside. An atmosphere
of calm and quiet, and of love of the
world while being removed from it,
permeates Quarton's painting.

Fig. 10: Hans Memling's triptych of *The Last Judgement*, *c*.1467-73. Memling was around forty years old when he painted this work. Born in Germany, he lived in Brussels and Bruges, and his paintings, which represent one of the summits of Early Netherlandish art, feature redheaded Virgins as well as angels and here, two souls of the saved, walking up the crystal stairs to Paradise.

Fig. 11: Studio of Martin Schongauer, *Noli Me Tangere*, *c.* 1480. Schongauer, like Memling, was a follower of Rogier van der Weyden. His works, like Memling's, are prized for their use of color and the quality of their execution. And like Memling, Schongauer created redheaded Virgins, saints, and angels, as well as Magdalenes.

THE EXCREMENT
OF THE HEAD

Each people has its own barbarians.
HERODOTUS, *HISTORIES*

L ondon, Christmas 1578, and the translator and editor Ralph
Holinshed is learning just how difficult it can be to please
all of the people all of the time.

In the best tradition of projects from hell, the one now causing
him such concern shouldn't even have been his. More than twenty-
five years before, Holinshed, then a stripling of twenty or so, had
been taken on as an assistant by one Reyner Wolfe, a Dutchman
who had arrived in London in 1533 and set himself up (with some
success) as a bookseller and printer. The publishing world was as
hungry for the next big thing then as it is now, and in 1548, a year
after the death of Henry VIII, and one year into the reign of his son,
the boy-king Edward VI, Wolfe had hit upon a plan for a publication
that would leave his rivals in the shade.

What Wolfe had come up with was the concept of a "Univer-
sal Cosmographie"—a kind of Tudor Wikipedia. This mighty
work would contain in two volumes a complete and newly written
account of the history of every nation then known, from the Biblical

Flood up to Wolfe's own present day. With what one can only view as heart-warming Anglophilia, Wolfe had decreed that volume one would be devoted solely to the history of Britain, while volume two would deal with the rest of the planet. Now, two and a half decades later, the project has finally been published. In the intervening years, King Edward VI has died at the age of fifteen and been succeeded by first his cousin Lady Jane Grey, the pitiful "Nine Days Queen," then by his half sister Princess Mary, or "Bloody Mary," as she became known. In 1558 Mary in turn had been followed to the throne by her half sister, Elizabeth I. And volume one of the "Cosmographie" had grown. The description and history of England, Scotland, and Ireland were now the entire work. And Wolfe too had died, in 1573, and his backers and his widow had handed the entire project, plus its contributors, over to Holinshed (with, one imagines, a mighty sigh of relief) to bring to completion. And astonishingly, twenty-five years after Reyner Wolfe first came up with the idea, Holinshed's *Chronicle*, as it has ever since been known—one of the largest books that has ever been printed in England—is a best-seller.

Just as well, too. Best guess is that the work took its printer a year and a half simply to typeset. We can imagine many a quill-penned cost sheet being argued over, many a candle being burned low into the small hours, but nonetheless demand and interest were enough for its backers to press on. By 1584 it was clear that the original edition would sell out (a highly desirable state of affairs for any publication, then or now), and the *Chronicle*'s backers would set about creating a new and even more extensive revised version, which Shakespeare would famously use as a source for *Macbeth, King Lear,* and for his history plays. But all that is in the future. Right now, the *Chronicle* and all those involved in it are looking at a significant problem. On December 5, 1578, the Queen's Privy Council bans all further sales of the book, and Richard Stanyhurst, one of Holinshed's writers on the *Chronicle*, is summoned to appear before them. Plainly, something among the book's two-and-a-half-million words has offended someone.

This was not quite such a heart-stopping prospect as it would have been during the reigns of Elizabeth's father or her sister, Mary, when such a summons was all too often the start of a short walk to imprisonment and the gallows, but it would nonetheless have been enough to give any publisher a sleepless night, with nightmares of bonfires of books on Tyburn Hill, and the publisher himself facing a hefty fine or even worse. The consequences could be ruinous.

And as Holinshed must have been all too aware, the *Chronicle* could not but, in places, sail close to the wind. It came right up to the (then) present day; thus it not only had to deal with the execution of Elizabeth's mother, Anne Boleyn, in 1536, it also had to reach some kind of a summation of the reign of her father (not to mention those of her half brother and half sister), when the dust was still settling from the social and religious upheaval of the Reformation, and with Protestant England still defining its position on the edge of a largely Catholic and more or less hostile Europe. With this the temper of the times, it's unsurprising that the lines around who was your friend and who were your enemies were drawn in thick and black. This is one reason why Holinshed's *Chronicle* is so much a part of our story here: its recording of Elizabethan England's attitudes toward its "other," in the shape of the Irish and the Scots. But before that there is its place in the iconography of one of the most famous redheads in history: Elizabeth herself.

One of the innovations of the 1577 *Chronicle* is that it includes illustrations—woodcuts—integrated at relevant points into the text. Thus we have Macbeth meeting the witches in the account of his reign, or in the "Description of Ireland," an Irish chieftain feasting his boisterous retinue out-of-doors. Many of the illustrations are charmingly anachronistic (Macbeth's witches wear a sort of Elizabethan masque costume; Macbeth himself is in ribbon breeches and a fashionable beaver hat), a fact that seems not to have troubled Holinshed at all. The images are there, so far as we can judge, simply to help the reader visualize what is described in the text, not to represent it. They also appear to have been pretty crudely made; in many cases the printer simply reused existing blocks. The artist or artists of those newly made for the work hasn't been identified, although Marcus Gheeraerts the Elder (*c.* 1520–*c.* 1590) has been suggested as their designer by some. Gheeraerts was noted as an etcher and printmaker and if he really did have a hand in the *Chronicle*'s woodcuts, there must have been a considerable distance between design and execution, because the illustrations in the *Chronicle* are definitely at the cheap and cheerful end of Tudor art. But then that is in the nature of woodcuts: reused, recut, and much repeated, they are the artistic equivalent of a game of telephone. Details get exaggerated, others are lost altogether. You can never be quite sure if what you see is what the artist originally intended.

For example, one of the most well-known portraits of Elizabeth, even among those for whom a copy of the *Chronicle* was out of reach, would have been the frontispiece to the *Queen's Prayer*

Fig. 14 The frontispiece to the Protestant *Queen's Prayer Book* of 1569, showing Elizabeth I at prayer, in a woodcut supposedly by Levina Teerlinc. These images took the place in the book of traditional illuminations. Their presence is telling evidence of how the Anglican Church was still feeling its way toward a public image.

Book of 1569, showing Elizabeth herself at prayer. This was supposedly cut by one Levina Teerlinc. Elizabeth, unlike her father, had no official artist (you can't help but suspect this was because she wanted no one in charge of the queen's image but the queen), but at this date Teerlinc was perhaps the closest thing to it, in which case the clumsiness of this image of the queen is all the more extraordinary. Elizabeth, the Virgin Queen, facing right, and neatly taking the place that would otherwise have been reserved for the Virgin Mary in what has been called a Protestant Book of Hours, is shown in profile, with trademark nose, and a forehead as high and bald as an Aztec priest (Fig. 14). It was fashionable to pluck the hairline, yes, but this is absurd. Yet whoever cut the image of Boudicca addressing her troops for the *Chronicle* had seen it, giving Boudicca the same profile (Fig. 15). They give her the loose, flowing hair of Herodotus's description of the queen of the Iceni, but also that of Elizabeth's own "Coronation portrait," where her loosened hair signifies her youth and virginal status, and they place a knopped crown, with open arches and very similar to the one Elizabeth wears in that painting, on Boudicca's head.[46] There is no attempt to represent Boudicca in Celtic robes, any more than the men-at-arms she is addressing are made to look like Celtic warriors. Boudicca wears what looks like a sumptuous embroidered Elizabethan gown, while her soldiers are armored as would have been any crack regiment in Elizabeth's own army. It's difficult to imagine any Elizabethan looking at this image and not seeing in it a reference to their queen and their armored men of war, defying not the might of Imperial Rome but that of Hapsburg Spain. England was as terrified of invasion in 1578 as it would be in 1940. And if it wasn't the Scots, threatening to let the French in down the chimney, it would be the Irish, letting the Spaniards in through the back door.[47]

46 The "Coronation portrait" of the queen can be seen today at the National Portrait Gallery in London, or on their excellent website, as NPG 5175. The painting is now thought to be a copy, made *c.* 1600, of a lost original of *c.* 1559. The cloth of gold Elizabeth wears in the portrait had also been worn by her half sister, Mary I. It's not so far removed from the brocade in which Enguerrand Quarton dresses his Queen of Heaven for her Coronation.

47 The whole of Europe knew how the Spanish treated those they saw as heretics, and while it's most unlikely anyone in England cared one way or the other how Spain dealt with its Jewish or Moorish *conversos*, let an honest Englishman find himself caught in the thralls of the Inquisition, as happened to one Robert Tomson in 1556, and the outrage was national. Tomson's account (another Tudor best-seller) was republished in G. R. G Conway's *An Englishman and the Mexican Inquisition, 1556-1560*, Sidewinder Studies in History and Sociology, 1997.

Fig. 15 Boudicca addressing her troops from Holinshed's *Chronicles* of 1577. Note the very Elizabethan-looking armor on her men-at-arms. The hare is being used as a means of prognostication; the little scene in the tent in the background seems to record the treatment of the queen and her daughters at the hands of the Romans. But to the Irish of this period, the invaders were the English themselves.

By 1578 Elizabeth I, also known as "Gloriana" and "the Virgin Queen," the archetype of a Protestant monarch, let alone of a Protestant female monarch, was in the twentieth year of her reign and the forty-fifth year of her life. This was way past one's prime as a woman in the sixteenth century, yet to look at Elizabeth's portraits, you would never know it. It may have been quietly accepted by her subjects that their queen would never marry, that she was beyond the age when there would be any Tudor heirs, just as it seems to have been apparent to all by this point that Protestant England was heading slowly and steadily into war with Europe's superpower, Catholic Spain. According to Elizabeth's image and her image-makers, however, none of this had happened. Time had stood still. The queen was still as young and wrinkle-free, as limber, as on her

accession day. Only the magnificence, if one may put it that way, of her presence had increased as the years had passed.

One of the earliest portraits of Elizabeth was a gift, it is thought, from her to her brother, Edward, and shows her as a girl of sixteen or so. Psychologically it is extremely astute: Elizabeth looks both tightly wound and incapable of being fooled by anything.[48] It shows her with smooth, gingerish hair, parted in the center (although more or less hidden under her Tudor gable-style headdress), and with her mother's dark brown eyes—sadly wary in their expression, in her daughter's case. Anne Boleyn had been a glossy brunette but obviously must have been carrying that redheaded gene; Elizabeth's red-haired father, Henry VIII, was, as Holinshed describes him, "in his latter days, somewhat gross, or as we terme it, *bourly*." Burly, that would be, but with a hint of other meanings to it, too: "boorish" being one, and boar-like another—short-tempered, unpredictable, highly dangerous, and like Chaucer's miller, with his inner nature announced by his red hair and bristles. Given that he had both beheaded her mother and declared her illegitimate, you might have thought Elizabeth would have been tempted to distance herself visually from her father, but no. Elizabeth's public image, once she became queen, is a celebration both of redheadedness and of the pale skin that so often goes with it, and it is an elective one. Once the gable headdress was out of fashion, Elizabeth's hair is as public as the rest of her—curled and tight to the head, almost boyish, when she first became queen, in keeping with the somewhat masculine

---

48 The portrait is part of the Royal Collection and hangs at Windsor Castle. See http://www.royalcollection.org.uk/collection/404444/elizabeth-i-when-a-princess.

fashion for tight doublets and high collars in the early Elizabethan period. Then, as Gloriana's wardrobe became ever more elaborate, those tight, tight curls ascend to heights that are almost Pharaonic (Fig. 16). Ordinary people did not have hair like this, any more than ordinary people wore foot-high collars to set off their head with the spread of gauzy wings. Of course they did not; they were not the queen. Unsurprisingly, most of these later hairstyles, glinting with diamonds and pearls, are wigs. Unless Elizabeth wished to spend half the day under the hands of her hairdresser, wigs were the only way to achieve such a coiffure, and in any case, how would natural hair support the weight of those jewels?[49]

Bewigged, therefore (she is said to have owned eighty), Elizabeth might have sported hair of any color she wished, yet she chose red—rarely a popular choice, and most likely more popular in England in her day than at any time up until our own. Red hair and pale skin were the Elizabethan brand, if you like, and those courtiers not blessed genetically, and all those many more wishing to copy the fashions of the ruling class, might dye their beards red if men (there is a splendid portrait by Marcus Gheeraerts the Younger of Elizabeth's favorite, the Earl of Essex, with a fashionably square-shaped, fox-red beard), or if women, change the color of their hair in order to emulate the queen with the use of such folksy tinctures as rhubarb juice, or the rather less attractive-sounding oil of vitriol (that would

49 Elizabeth's use of wigs has prompted speculation that she lost her hair when she fell ill with smallpox in 1562. This seems to be based on an errant bit of historical supposition started by F. C. Chamberlain in 1922. Elizabeth did however go gray, if the evidence of the lock of her hair preserved at Wilton House is trusted. Redheads "go gray" just as much as any other hair color, but if you're lucky, with red hair the unpigmented hairs are to an extent disguised, and simply look like fairer hair among the red.

be sulphuric acid to the rest of us). Elizabeth is even, at one time, said to have dyed the tails of her horses orange.[50] For the pale skin there was of course white lead, splendid for giving the skin a satiny white finish, and horribly injurious to health in any degree of contact whatsoever. We can hardly assume the Elizabethan court was ignorant of this, as the hair fell out and the skin withered, and the headaches and tremors took possession of the first ladies-in-waiting to succumb to lead poisoning. But for Elizabeth, the use of such cosmetics may have been a necessity. She may not have naturally enjoyed the ethereal, moonlike pallor with which she is depicted, and which so often accompanies red hair. In 1557 the Venetian ambassador Giovanni Michieli described the queen-in-waiting as having "good skin, although swarthy." Another Italian diplomat, Francesco Gradenigo, describes her in 1596 as "ruddy in complexion." Possibly Elizabeth was sensitive about her apparent high color; Sherrow's *Encyclopedia of Hair* records her as having asked whether her hair was superior to that of her cousin and rival, Mary, Queen of Scots (in 1578, languishing in the eleventh year of captivity in England), and if her skin was fairer. In her happier youth, Mary, Queen of Scots, had been married to Francis II of France, and there is a memorable portrait of her painted by the French artist Francois Clouet in 1560, showing her in all-white mourning following the deaths of her own mother and of her father-in-law, the French king Henry II. It shows

---

50 Victoria Sherrow, *Encyclopedia of Hair: A Cultural History* (Westport, CT: Greenwood, 2006). It didn't do to emulate the queen too successfully, however. Lettice Knollys had married another of Elizabeth's favorites, Robert Dudley, in September 1578, without royal permission. To judge from her portrait of *c.* 1585 by George Gower, Lettice was as head-turning a redhead as Elizabeth (portraits of the two can be extremely difficult to tell apart), and she was ten years younger. Lettice was banished from court, never to return.

her with skin almost as pale as her mourning veil and hair of a very dark red. However on her execution in 1587 at the age of forty-five, it was discovered that Mary too had resorted to wearing a wig, and that her natural hair was by then a close-cropped gray—"as gray as one of threescore and ten years old," in the words of Robert Wynkfielde, a witness to her death. You wonder how acutely aware these two women, cousins and once sister queens of sister kingdoms, might have been of each other's looks.

But why would Elizabeth create an image so frankly outlandish and dangerous to maintain? Part of the explanation might lie with Elizabeth's own complex psychology. She had been declared illegitimate by her father; therefore to parade his red hair so prominently was one way of giving the lie to that. Her mother had been beheaded and Elizabeth herself, during the reign of her sister, Mary, came close to the same fate. In Elizabethan portraiture there is a great deal of attention devoted to the head and the surroundings and the accoutrements of the head—jewels in the hair, earrings, the framing element of the famous Elizabethan ruffs of the 1580s, which put the head almost on a platter, John the Baptist–style; and then the enormous stand-away, look-at-me collars of the 1590s.

Another part of the reason may lie within the lines of the "Hymn to Astraea," a grandiloquently awful poem written for Elizabeth in 1599 by Sir John Davies, to celebrate the anniversary of her accession.[51] Davies was an Elizabethan heavyweight in every sense, and his flattery of the queen is not so much laid on as ladled:

> *But here are colours, red and white*
> *Each line and each proportion right;*
> *These lines, this red and whitenesse*
> *Have wanting yet a life and light*
> *A Majestie and brightnesse . . .*

And so it goes on. There are twenty-six of these hymns, and the first letters of each line in each of them do indeed form an acrostic of the queen's name. But the colors Davies singles out as "being" Elizabeth—red and white—are also the colors not only of the Tudor rose, the imprimatur of the Tudor dynasty, but equally of the flag of St. George, the patron saint of England (as he is to this day) and the only saint whose banner remained in use in England after the break with Rome. The poet Edmund Spenser's epic *The Faerie Queene*, of 1590, would use the same badge of a "bloudie crosse" to distinguish its hero, the Redcrosse Knight, from the various personifications of foreign villainy that pop up throughout the poem. It is, admittedly, hard to read this work today without Monty Python's "Knights who say *Ni!*" creeping into your thoughts, but it still does a pretty good job of conjuring up Elizabethan England's xenophobia. And its queen's royal branding, in red and white (remembering also that white was the color of virginity, for a Virgin Queen), and

---

51 The portly Sir John died of apoplexy in December 1626, after what must have been an extremely good supper-party celebrating his appointment as Lord Chief Justice. He had been lobbying for the role for years. His wife, Dame Eleanor Davys, believed herself to be a prophetess, giving rise to the anagram *Never so mad a ladye.* But she foresaw the date of her husband's death and wore mourning for the three years before it came to pass.

the magnificence of her image-making, created an aligning of sovereign, symbol, and state of a different order to anything that had gone before. Ironically, in later portraits of Elizabeth, with a few rare exceptions, the splendid clothes and the jewels may be the only elements painted from life—the ivory face and crown of red hair are an icon, mass-produced, and instantly recognizable even today. The "Elizabeth I" brand is by that measure one of the most long-lived and successful in history.

Only toward the end of her life, when she was poignantly described by Sir Walter Raleigh as "a lady whom time has surprised," does Elizabeth seem to have varied the formula, with wigs just as high but paler, fairer in color: "her hair," so the German lawyer Paul Hentzner describes her, at Greenwich in 1598, "an auburn color, but false." In this case Hentzner seems to have been using the word "auburn" in its original sense of brownish-white. The term that would later become so handy and socially acceptable a catch-all for hair with any reddish tint to it at all first entered the English language around 1430, and comes from the Latin *alburnus*, or white. Only in the seventeenth century did its sense change, and the notion of it meaning a color more brown than pale become common. Ever since then it has been red hair's aristocratic first cousin, but the color Hentzner meant is probably that shown in Robert Peake's *Procession Picture* of the queen of 1601 (Fig. 17). Perhaps the pallor of seventy-year-old skin prompted the change, when the contrast with a red wig began to look too harsh. Elizabeth understood her coloring. Even today, any redheads wishing to know which colors they should favor in their wardrobe need look no further than her portraits.

The *Chronicle*, then, was created at a time when there was a sense of Englishness as being more than nationality, of it standing for some nebulous but estimable set of qualities that governed not merely a man's (or woman's) tongue but his or her beliefs and attitudes and values. It contains something of everything Elizabethan England knew or believed of itself. It is, if you like, a stethoscope, laid against the Tudor heart. Let the pope excommunicate us all (Elizabeth herself had been excommunicated in 1570); we knew better—not for nothing was the so-called Act of Supremacy so called.[52] There was no more Mary Magdalene to bring your woes to, no more pope, no more Church of Rome. Not for nothing, either, had a work that was to have recorded the whole world become one that devoted itself instead to describing one rainy island for the delectation of its subjects. Here too Holinshed caught the temper of the times: the narrative arc of the *Chronicle* is upward to the (then) present day, the "perfect monarchie," the happy ending represented by Elizabeth I—if, that is, this halcyon moment, this little earthly paradise, could only be preserved from its enemies, both abroad and those rather nearer home.

Holinshed's writer for the Scottish chapters of his *Chronicle* was one William Harrison. Harrison's life epitomizes the seesawing nature

---

52 The 1558 Act of Supremacy finally established the Protestant faith as the official religion of the state and made the monarch, instead of the pope, head of the Church of England.

of what it was to be English during this period. Born in 1534, he had been raised a Protestant; converted to Catholicism while a student at Oxford during the reign of Mary I; converted back again slightly before her death, and by 1559 was establishment enough to have been granted his first living as a clergyman. Harrison writes rather daintily, as if he were in conversation with his readers. For him, where Scotland was concerned, it was a question of united we stand:

> If the kingdoms of Britain had such grace given them from above as they might once live in unity, or by any means be brought under the subjection of one Prince, they should ere long feel such a favour in this amity that they would not only live frankly on their own without any foreign purchase of things, but also resist all outward invasion, with small travail and less damage. . . . [53]

He sounds like Herodotus bemoaning the state of Thrace, and it shouldn't be surprising that he does. Scotland and Ireland were as much England's "other" as Thrace had been to Greece. And there is that fear of invasion again. Sadly for Harrison, there was not much to be expected of England's closest neighbor; for him, the abiding characteristic of the Scots is that they are drunks. Although he describes them as otherwise "courageous and hardy. . . . They cannot refrain from the immoderate use of wine," with the result that "if you knew them when they be children and young men, you shall hardly remember them when they be old and aged . . . but rather suppose them to be changelings and monsters." These are the lowland Scots.

---

[53] The Holinshed Project gives the text of both editions of this work, and just about every tool a researcher could ask for to search through them. See http://www.english.ox.ac.uk/holinshed.

Highlanders fare rather better, being described as "less delicate and not so much corrupted by strange blood and alliance" (that strange alliance would be the one with the French). They are "more hard of constitution . . . watch better and abstain long . . . bold, nimble and more skillful in wars"—all qualities, we may reflect, that would be sorely needed by the Scots and sorely tested in the following centuries of English rule. The one thing Harrison does not do, however, which may seem unexpected, is castigate the Scots on the basis of their hair color.

In fact there is little in the way of description of the appearance of peoples or individuals of any sort in the *Chronicle*. William Rufus is described as "William the Red," with some pointed reflections on his character. King Henry II of England, invader of Ireland in 1171, has his "rednesse" recorded for posterity (red hair and ruddy skin; very suitable for such a warmonger of a king), but beyond those two, not much. There might have been more in the description of Boudicca, but the *Chronicle*, while leaving one in no doubt of the impression a queen might create upon her subjects—"Her mightie tall personage, comely shape, severe countenance and sharp voice . . . her brave and gorgeous apparel also caused the people to have her in great reverence"—has her as a blonde, with "her long and yellow tresses of hair reaching down to her thighs." A matter of translation, perhaps, or you might wonder whether this wasn't simply being politic. With an actual red-haired queen on the throne, it would hardly increase your success as a publisher to describe her hair color as being a characteristic of the treacherous barbarian, or even worse, of a queen who lost her life to an invading force. However, there is a third possibility. The evidence is scanty and very circumstantial, but it is intriguing. It might just be that red hair in the sixteenth century was less common in Scotland than it is today.

John Munro, he of those opinions on the coloring of the Vikings, also writes in 1899 of the Scots as being "of all complexions, from very dark to very fair, with a dash of red hair, about 4–5 percent more or less, in different localities, that is to say rather more than in England." That is to say, significantly less than the 13 percent of the population estimated to sport red hair in Scotland today. This would rather give the lie to all those tedious reports of redhead extinction. While in 1911 the Irish journalist T. W. Rolleston, in his *Myths and Legends of the Celtic Race*, would write that "the prevalence of red hair among the Celtic-speaking people is, it seems to me, a most striking characteristic . . . eleven men out of every hundred whose hair is absolutely red," suggesting that the present proportion of redheads in Ireland hasn't reduced by a single percentage point in the past hundred years, either.

Perhaps the number of redheads in Ireland in the sixteenth century was also lower than it is today. Richard Stanyhurst (in December 1578 awaiting that nerve-racking interview with Elizabeth's Privy Council) describes the Irish thus. First he accuses them of having the Spanish as their "mightiest ancestors," therefore being pretty much born traitors. Then, he says, they are uncouth, they live like animals, and they corrupt the English language—making a "mingle-mangle" of it, in Stanyhurst's own memorable phrase—and without constant vigilance do the same to the manners and morals of any of the English sent to dwell among them.[54] (The poet Edmund Spenser was even less pleasant, warning against the use of Irish wet nurses, as if barbarism was something that might be sucked in with a mother's milk.) The description goes on; the Irish are "religious, frank, amorous, irefull . . . very glorious, many sorcerors, excellent horsemen, delighted with wars . . . the men clean of skin and hew, of stature tall, the women well-favored" and "proud of their long, crisped bushes of hair." But there is not a mention of that hair's color as being red. And Stanyhurst should have known what he was talking about; he, like Edmund Spenser, was Anglo-Irish, one of those many whose ancestors had been shipped in from England or Scotland to occupy lands around Dublin, in a piece of thickheaded imperialism that unworked in Ireland in the same way as it would on the West Bank four hundred years later. English foreign policy in Ireland was a disaster, had been for centuries. No wonder the Privy Council was so sensitive about it. No wonder they thundered on about "report of matters that . . . are not meet to be published in such sort" being put out there for discussion in a book that "falsely recorded events." One suspects rather that Stanyhurst unwittingly recorded events all too well.

Or we might simply need to adjust our vision and read the *Chronicle* as an Elizabethan would have done. Red hair might not be made much of in its pages, but the Scythians and their "red haire" certainly were. This is the description from the *Chronicle* of the first

---

54 Stanyhurst could give James Joyce a run for his money. Among other gems in the *Chronicle*, he comes up with the phrase "idle benchwhistlers" for the lazy and describes those who would take credit for another's work as flies who fall in another man's soup. Rather less likeably, he also describes the Irish language as "a ringworm."

inhabitants of Britain:

> The people called Picts invaded this land, who are judged to be descended of the nation of the Scythians, near kinsmen to the Goths, both by country and manners, a cruel kind of men and much given to the wars. This people . . . entering the Ocean sea after the manner of rovers, arrived on the coasts of Ireland, where they required of the Scots new seats to inhabit in: for the Scots which (as some think) were also descended of the Scythians did as then inhabit in Ireland. . . .

And this is what an Elizabethan understood of the Scythians. This is from *King Lear*, which, with *Macbeth*, could claim to be the *Chronicle*'s greatest legacy:

> *The barbarous Scythian*
> *Or he that makes his generation messes*
> *To gorge his appetite, shall to my bosom*
> *Be as well neighbour'd, pitied, and relieved,*
> *As thou my sometime daughter.*

And here is Edmund Spenser, again going even further: "I Suppose," he says, the Irish "to be Scithians." They are "the most barbarous Nation in Christendom," Spanish blood (the ancestors of the Irish, remember, along with the Scythians) "is the most mingled, most uncertain, and most bastardly," and as for Scotland, it and the Irish are "one and the same."[55] It is deeply depressing to read the same piece of nonsense linking the Irish, the Scots, and the Scyth-

---

55 *A View of the Present State of Ireland*, 1596. This pamphlet is so unpleasant and so incendiary it was kept secret during Spenser's lifetime. Spenser bought lands and an estate in Ireland; he was also present at the Smerwick Massacre of 1580, when a ridiculously small force of maybe five hundred Spanish and Italian troops first surrendered to and were then killed by English soldiers.

ians being put forward by John Munro three hundred years later. Tying himself up in knots to account for all those random redheads, Munro decides that immigrants from Scythia first populate Ireland, then leave it, after exactly 216 years, to populate Thrace, then return to Ireland and after that people Scotland, too. Perhaps one shouldn't put too much trust in his math, either.

Might there have been fewer redheads in Scotland and Ireland then than there are now? Given the workings of genetic drift, it's certainly possible. But whether red hair, rather than general barbarity, was associated with either nationality at this date, the reviling and denigrating of the Scots and Irish by the English is beyond doubt. Edmund Spenser would have cleared the Irish from Ireland entirely—language, culture, customs, and people. The *Chronicle* played its part in enshrining their outsider status, and the red hair of the Celts inherited and suffered under these same attitudes as well. One red-haired queen did not redress the balance.

"A conquest," in Stanyhurst's view, "ought to draw with it three things, to wit, law, apparel, and language." Neither conquest nor union with Ireland or Scotland drew with it from England any such things. What Scotland and Ireland, misprized for centuries, drew from their relationship with England goes on to become a part of the history of red hair in the New World as much as in the Old. As for Stanyhurst, he survived his interview with the Privy Council, but on emerging from it he promptly left the country, never to return. Some years later he was working in the alchemical laboratory at the Escorial of Philip II of Spain, the king who was to launch the Armada against England in 1588. One wonders if Stanyhurst shared with the king his theories on Spain's links to those Irish barbarians. He ended his days as chaplain to the Catholic Archduke Albert of Austria in the Netherlands.

Holinshed, meanwhile, the poor soul, having patiently excised all those pages from his *Chronicle* that had so provoked the wrath of the Privy Council, retired to the country and died two years later.

Sometime after Shakespeare's death in 1616, *Macbeth* was revised by Shakespeare's near contemporary, the playwright Thomas Middleton, and among his additions was a song, "Black Spirits." This seems to have been lifted from Middleton's own play *The Witch* of 1615. By now there was a Scottish king, James I and VI (son of Mary, Queen of Scots), on England's throne, who was known for his fascination with witchcraft.[56] Black magic was, well, the new black.

The lines Middleton inserted into Shakespeare's *Macbeth* include the usual Addams Family list of ingredients: blood of a bat, libbard's (leopard's) bane, juice of a toad, oil of an adder, and "three ounces of the red-haired wench." Scholars have speculated that the red hair is to be read as an allusion to lechery, or as an allusion to Jewishness and the Jews' anti-Christian rites, or as an allusion to poisonous substances, or as all three.[57] But this is *Macbeth*, the Scottish play. It might simply be a rare touch of local color. It might also be playing with

---

[56] James inherited the Scottish throne, as the sixth king of that name, from his mother, Mary, Queen of Scots. He was also a great-grandson of Henry VIII's sister Margaret and inherited the English throne, as England's first King James, from Elizabeth on her death.

[57] Jeffrey Kahan, "Red Hair as a Sign of Jewry in Middleton's Additions to *Macbeth*," *English Language Notes* 40, no. 1 (September 2002).

another ancient prejudice directed at red hair—that it is somehow connected to witchcraft and the supernatural.

The figure of red-haired Lilith today has abandoned her Babylonian wastes for a shabby hinterland somewhere between mythology and pornography, but originally conflated with the *striges*—vampirelike demons of Ancient Greece—she killed children in their cots and seduced men in their sleep. Wet dreams were Lilith's doing, and she might also leave her victims impotent or even cause their penises to disappear entirely (still an accusation thrown at those suspected of witchcraft in Africa today). Woman as sexual predator has always terrified and aroused in equal measure, and witches have always been bewitching, in art and popular culture at least. The reality was rather different.

The handbook in the early modern age for witchery and witchfinders both was the *Malleus Maleficarum*, written in 1486 by one Heinrich Kramer, aide to the Archbishop of Salzburg, with another German clergyman, Jacob Sprenger, as a kind of publicist and PR man for the work. The pair of them were charlatans to a degree unrivaled even by Aleister Crowley and Montague Summers in the twentieth century. Unsurprisingly, Summers was the work's translator into English. Thanks to him, to this day you will find the *Malleus Maleficarum*, or *Hammer of Witches* as it is known, cited as an authority for tales of young, nubile, redheaded, green-eyed women being dragged off to horrible deaths at the stake. But if you read his translation, or as much of it as you can bear, what you find is a work of unrelenting misogyny that held all women in equal contempt, whatever their hair color might have been. Witches, it says, are driven by lust, and those most likely to be witches are adulteresses, fornicators, and concubines; and those most likely to be victims of the unwanted attentions of the devil are "women and girls with beautiful hair; either because they devote themselves too much to the care and adornment of their hair, or because they are boastfully vain about it." It certainly doesn't have to be red.

In reality it was pretty much certain to be gray or even white. While desirable young women may be depicted in the art of the period as witches, as in the works of Hans Baldung Grien in particular, and even with the *Chronique de France* of 1492 recording the Frankish kings as burning red-haired women as witches seven hundred years before, in the great witch hunts in Europe of the sixteenth and seventeenth centuries, those going to the stake or the gallows on a charge of witchcraft were almost bound to be poor, elderly, widowed, and unprotected.[58] This was recognized even at the time. Reginald Scot (*c.* 1538–99) writes in his *Discoverie of Witches*, a work that bravely set out to prove there was no such thing, that "one sort of such as are said to be witches, are women which are commonly old, lame, bleare-eyed, pale, foul, and full of wrinkles, poor, sullen, superstitious or papists, or such as know no religion, in whose drowsy minds the devil hath gotten a fine seat."

The beauteous redheaded victims of European witch hunts exist in our imaginations only, and in so imagining them, we are being seduced ourselves by an association between otherness and otherworldliness, between red hair and supernatural forces, and between red hair and erotic circumstance that simply refuses to quit: there

---

58 Brian P. Levack, *The Witch-Hunt in Early Modern Europe* (Oxford: Routledge, 2006).

was red-haired Malachai in Stephen King's *Children of the Corn* in 1984; there is the sorceress Melisandre, the Red Woman of *Game of Thrones*; and in 2014 the casting director of *The Last Witch Hunter* was advertising for a redhead with a pale complexion as the movie went into production.[59] How very confusing all this can be.

Or as Obadiah Walker put it, "Each man disparageth his fellow-creature, and gratifies his haughty humor in the derision of his brother." Obadiah Walker was Master of University College, Oxford, from 1676 to 1688, where supposedly his melancholy ghost still walks. He lost his post for refusing to abandon his Catholic faith, so knew more about the workings of discrimination than many another. In his *Periamma epidemion, or, Vulgar errours in practice censured* of 1659, Obadiah writes of "a common yet causeless calumniation: viz. the vilifying of red-hair'd men, the putting of disesteem upon persons, merely because of the native color of the excrement of the head." He means the hair, but you do begin to think sometimes that the phrase might as justly be applied to the historic bigotry and prejudice in our thinking. "I could wish," he says "that men would not hoodwink themselves with their own prejudice." But they do so still. As far as that as a signifier of barbarity goes, it's global.

And yet, and yet. . . . If "other" repels, it also fascinates. There

---

59 In 1887, in the *Ancient Legends, Mystic Charms, and Superstitions of Ireland*, you might read this: "Red hair is supposed to have a most malign influence, and has even passed into a proverb—'Let not the eye of a red-haired woman rest upon you.'" Who is the writer? Some chauvinist Englishman? Some witch-hunting mittel-European cleric? No, it is Lady Francesca Speranza Wilde, Irish mother of Oscar. More recently, in her thesis *Sirens and Scapegoats: The Gendered Rhetoric of Red Hair*, Emily Cameron Walker draws attention to the number of times the disgraced CEO of News International Rebekah Brooks was referred to in the English press as a red-haired witch during her trial in 2014. See http://ecameronwalker.blogspot.com/2012/09/thesis.html.

is always that contradictory desire within us to stand out. Even in the eighteenth century, when any hair color other than gray was as unfashionable as could be, when every head wore a wig, and every wig looked like powdered topiary, there was a moment in 1782, recorded by the diarist John Crozier, when despite "much aversion as people in general have to red hair, the appearance thereof was so much admired that it became the fashion, for all the Beaus and Belles wore red powder."[60] Nor is the fashion for red hair restricted to the West. Something very similar to the craze for red hair in London in the 1780s happened in Japan at the end of the twentieth century, and the Japanese had otherwise reviled red hair for centuries.

When the first Westerners set foot in Japan in 1543 (the same year, incidentally, that the nine-month-old Mary Stuart was crowned Queen of Scots) the Japanese were appalled by them. These crude, semi-civilized creatures, who ate with their fingers and had all the self-control of children, were immediately compared by the genteel and sophisticated Japanese to monkeys and monsters, to legends of primitive wildmen, covered in fur, and all such, without any reference to their actual appearance, had been lumped together under the label of "red-haired barbarians." Body hair and hair on the head in any color other than Japanese black became the dominant symbol of otherness in Japan for centuries.[61] Yet in a society where the term *ang-mo*, or "red-haired ape," is still bandied around as an insult to

60 Quoted in C. Willet Cunnington and P. E. Cunnington, *Handbook of English Costume in the Eighteenth Century* (London: Faber, 1957): 1952–9.

61 Alf Hiltebeitel and Barbara D. Miller, *Hair: Its Power and Meaning in Asian Cultures* (New York: SUNY Press, 1998).

foreigners, in the 1990s Japanese teenagers began dying their hair all shades of red and brown.

This fashion for *chapatsu*, or "tea-color" hair, became a national controversy. Questions were asked in Parliament. Schools created "hair police," and even now students with naturally brown or curly hair can be asked to prove it should not be naturally black and straight.[62] Japan is a conservative and very homogenous society, and some of its teenage fashions consequently can seem over the top (an accompanying fad was for glittery stickers of fake tears pasted to many tea-haired teens' cheeks). But we are all barbarians to someone. In the 1930s, *Japanese* hairiness became a major feature of the anti-Japanese propaganda coming out of China. As Professor Alf Hiltebeitel puts it, "Nothing is ordinary about hair. It gets into everything, but whatever it gets into, it never seems to be explained in the same way; rather it always seems to be used differently to explain something else." And queen or commoner, we all want to shake a brighter tailfeather than the rest.

---

62 See http://www.japantimes.co.jp/community/2013/07/29/issues/prove-youre-japanese-when-being-bicultural-can-be-a-burden.

SINNERS AND STUNNERS

The truth about red hair, like many other truths,
lies enclosed in a nutshell, generally a hard one, and
people are often very short of crackers.
THE PHILOSOPHY OF RED HAIR, 1890

The area around St. Paul's Cathedral in London is one of the few parts of the city where you can still summon up its past. Close your eyes and ignore the traffic; imagine instead of honking taxis the shouts of irate draymen, the creak and squeak of wooden wheels, the clop of hooves, the endless music of an endless press of people—probably very similar to those who throng the same streets today, if rather less well soaped and washed. The street names still record their presence, these hordes of ghostly Londoners, and their doings here: Pilgrim Lane, Ironmongers Lane, Limeburner Lane, Old Jewry. Watling Street, leading to St. Paul's, was trodden by Roman legionaries and then by Boudicca's rampaging army; close by, off Cheapside, were once to be found the colorfully named Pissing Alley and, even better, Gropecunt Lane, until times changed and renaming became inevitable (the rather more acceptable Love Lane, where no doubt exactly the same activity took place, still exists). Paternoster Square, Amen Corner, and Ave

Maria Lane celebrate the permanence of human faith and worship in this area, no matter what religion, and as Ave Maria Lane runs north to Newgate Street its name changes, to Warwick Lane. And in a house on Warwick Lane in 1865 there lived a girl of about sixteen whose name was Alice Wilding. She earned her living (or rather contributed to the household, which included her grandmother, two uncles, and at least one infant) as a dressmaker. And secretly, like so many girls of any age or class, she dreamed of a career on stage. Alice, however, was no head-in-the-clouds innocent: when, one evening early in the year, she found herself being first stared at and then followed up the Strand by a short, tubby, balding man of middle age and clearly a higher social class than hers, she seems to have taken the experience in her stride. Very possibly, this was not the first time such a thing had happened to her. Alice had a face both strong and feminine, feline eyes, milky skin, and the most wonderful head of red hair, something between copper and the color of a marigold. The man introduced himself as an artist, Dante Gabriel Rossetti, and in the account of the meeting recorded by his studio assistant, went on to explain that he was "painting a picture and her face was the very one that he required for the subject he was at work on." The man begged Alice to come to his studio at Cheyne Walk in Chelsea the next day, and to sit for him, promising her that she would be paid. Once satisfied that she understood, and had agreed, he went on his way. The next day, "Rossetti made every preparation to receive her and make a study of her head for *The Blessed Damozel*. His palette was set, the canvas on the easel and everything in readiness. . . . "

Alice stood him up. Of course she did. Go sit in an "artist's studio," for an artist she had very possibly never heard of, a man, on her own? Was he mad?

Fortunately for us, however, that wasted day in the studio is not the end of the story.

One significant development in artistic life in the nineteenth century is that we start to learn the names and in some cases details of the biographies of the artists' most significant models. Sir Frederic Leighton, president of the Royal Academy from 1878 to 1896, and arguably England's premier and most successful painter of the period, had an entire family of sisters, the Pullans, who sat to him and his circle. Rossetti had first Lizzie Siddal and later Fanny Cornforth, with Alice Wilding modeling for him as well. Most of these women were working class, and their relationship with "their" artists was one of social and economic dependency as well as collaboration: outside the artist's studio many, in fact all those who posed nude, would be regarded as little better than prostitutes (Fanny Cornforth, who became Rossetti's live-in housekeeper and whose robust humor and working-class attitudes alarmed his family and friends all his life, may very likely at one time have been a streetwalker for real.) Various ruses were adopted by these young women to cover for the true nature of what they did in the artist's studio. Ada Pullan listed herself on the 1881 census as an "art student," for example. It was just about feasible by the latter half of the nineteenth century for a woman to have a career in the arts, to enjoy some measure of financial freedom without having been born to riches, to hold some control over her own life in a way that would have been exceptional in previous centuries, and to negotiate a place for herself within an artist's circle that was neither wholly sexual nor without respect. One who managed to do so was Joanna Hiffernan, an Irishwoman who met the American artist James Abbott McNeill Whistler in a studio in Rathbone Place, London, in 1860, when she too would have been about sixteen. She went on to have a six-year relationship with him, acted in loco parentis to his illegitimate son even after their relationship as lovers had come to an end, and was the inspiration for some of Whistler's most sophisticated and innovative works, including his *Symphony in White, No. 1* of 1862 (Fig. 18), and *Symphony in White, No. 2* of 1864–5. Both paintings make wonderful play of her pale skin, soulful eyes, and almost oversize features, and her dark red hair, hair that Whistler would describe ecstatically as "a red not golden but copper—as Venetian as a dream."[63] Whistler's biographer Joseph Pennell described the woman herself as being not only beautiful but intelligent and sympathetic. Joanna was also someone who steered her own very independent course through the world. She was unconventional (in her morality) and daring (in her professional life), both qualities the world ascribes all too willingly to redheads, and she has a unique role in the history of art in that her most famous supposed "portrait," a work with which she has been intimately associated for decades, is one that as a redhead she simply cannot have sat for at all.

The story begins with a trip to Paris, where while modeling for the second *Symphony in White*, Jo, as she was known, met the French artist Gustave Courbet. In 1865–6 Courbet painted her portrait as *La Belle Irlandaise*, showing her in close-up, before a mirror, combing out a tangle in her hair and creating his own slab-sided, meaty

---

63 Quoted in Margaret F. MacDonald and Patricia de Montfort, *An American in London: Whistler and the Thames* (London: Philip Wilson Publishers Ltd), 2013, which contains much useful information on Whistler and his relationship to Johanna Hiffernan or Heffernan, as she was also known.

version of the ethereal beauty who had captivated Whistler. In 1866 she began an affair with Courbet and posed as one of the two women in his *The Sleepers* (Fig. 19). This painting, which was the subject of a police report on what seem to have been the only occasion in the nineteenth century when it was publicly exhibited, in 1872, shows two women, one dark-haired, one Jo, curled about each other naked in bed, supposedly sleeping after making love. It has sometimes been hailed as a ground-breaking depiction of lesbianism, but it is surely rather a male heterosexual fantasy about female lovemaking. The poses in which the two women are supposed to have fallen asleep are unnatural (and, speaking as an ex–life model, look excruciatingly uncomfortable, too). Jo's head seems oddly unsupported, and the expression on her face suggests extreme concentration in holding a difficult pose rather than languorous afterglow. Like much erotica, perhaps it hasn't aged very well. But it created an association between the Bohemian, free-spirited Jo and Courbet's erotic paintings that is running to this day.

Courbet's other notorious erotic work, also of 1866, is known simply as *L'Origine du Monde*. For those who don't know it, Courbet's "origin of the world" is, predictably, a close-up of the view the artist would have had if he had set up his easel at the foot of his model's couch and asked his model to raise her shift above her breast and to open her legs. There's a piece of Anglo-Saxon, no doubt familiar to many an historic wanderer in London's St. Paul's, that would describe the subject of the painting perfectly. Its framing excludes everything else, including head, arms, and lower legs. It was commissioned by one Halil Bey, an Ottoman diplomat who also owned *The Sleepers*, and who can therefore claim front rank among the grand dirty old men of erotic art, and the painting still takes one aback, even today. But there is simply no possibility of it being a portrait of Jo, untrammeled by society's mores as she may well have been, and saddled with all the sexual baggage of being a redhead as she undoubtedly is. The pubic hair of the woman in *L'Origine* is so dark as to be almost black. The pubic hair of a redhead is, unsurprisingly, red. In fact you can see a tiny suggestion of Jo's own pubic hair in *The Sleepers*, a minute triangle of gold above the leg of her bedmate. This second woman, an unnamed brunette whose dark hair is spread out across the pillow beside Jo's coppery curls, may have her sharp features echoed in another painting, recently discovered, of a woman's head, mouth open, dark hair thrown back and purporting to show *L'Origine*'s missing upper half. Or, the latter may have nothing to do with the former at all. But neither is Joanna Hiffernan.

Back to the Strand in 1865, and Rossetti's studio assistant, Henry Treffry Dunn, continues the story of Rossetti's vanishing redhead:

> Days and weeks went by, and he [Rossetti] had given up all hope of seeing the young lady again and had even abandoned the picture, when one afternoon in company with [the sometime art dealer Charles Augustus] Howell in the same part of the Strand, he again caught sight of her. He was then in a cab, [&] telling Howell what he was going to do he stopped the Hansom at a side street, got out and darted after the girl and at last overtook her. He reminded her of the promise she had given him and told her of his disap-

pointment at her not coming and at last persuaded her to enter the cab and drive with him to Cheyne Walk.[64]

You have to wonder at the change of heart. Perhaps the encounter was less alarming the second time around; perhaps this had begun to feel like Fate; perhaps by now Alice knew who Rossetti was; perhaps there had also been a few hours in front of her looking-glass, wondering why she had been given this face and hair if to do nothing with it? In Rossetti's case, the explanation of his persistence is simpler—he was, in the words of the novelist Elizabeth Gaskell, "hair mad":

> If a particular kind of reddish-brown crepe wavy hair came in he was away in a moment, struggling for an introduction to the owner of said head of hair. . . .[65]

Thus Rossetti is forever immortalized as a classic example of Man with a Thing for Redheads. Aside from Rossetti and his obsession, however, why are there so many Pre-Raphaelite redheads? The term has become virtually synonymous, just as "Titian red" would be later in the century. They are there in works by Frederick Sandys, whose gloriously rufescent partner, the actress Mary Emma Jones, modeled for him first as the Magdalene in 1862, and then as Perdita, Proud Maisie, Helen of Troy, and countless other red-haired icons throughout his life. Arthur Hughes used his wife, Tryphena Foord, as his model for *April Love* (1856) and *The Long Engagement* (1854–9),

---

64 Jennifer J. Lee, "Venus Imaginaria: Reflections on Alexa Wilding, Her Life, and Her Role as Muse in the Works of Dante Gabriel Rossetti" (msaster's thesis, University of Maryland, 2006).

65 Quoted in Henrietta Garnett, *Wives and Stunners: The Pre-Raphaelites and Their Muses* (New York: Pan Macmillan, 2013).

both works depicting unhappy lovers, and both making much play of Tryphena's ghostly skin and shining red-gold hair. Henry Wallis depicts the eighteenth-century poet and suicide Thomas Chatterton on his deathbed, in his garret, neglected poetry torn to pieces on the floor beside his lifeless hand. Chatterton's bright red hair leaps out from the painting's chilly gray and green palette and is perhaps used by the artist to allude both to his subject's sensitivity and to his passionate poetic spirit. John Collier, a late Pre-Raphaelite, creates an irresistible red-haired *Lilith* in 1887, wearing her symbolic serpent like a feather boa, and paints Lady Godiva as a redhead in 1898 (although seeming to miss the idea that her hair should be thick and long enough to hide her completely). Rossetti surrounded himself with redheads, both male and female. Dunn's successor as his assistant was the redheaded Manxman and later novelist Hall Caine, a man of such unusual yet engaging appearance he might have been designed by Mother Nature to play Merlin. And for a year from 1862, Rossetti shared his house with the flame-haired poet and all-round oddity Algernon Charles Swinburne (Fig. 20), of whom more in the next chapter. Joanna Hiffernan and Whistler were guests at the house in 1863; one wonders if they were entertained by Swinburne, whose party piece consisted of sliding nude down the bannisters, and who reportedly infuriated Rossetti by dancing all over his studio "like a wild cat."[66] And famously, Rossetti's first muse and eventually his wife had been the red-haired beauty Elizabeth Siddal, the model for Millais's *Ophelia* of 1852, who was described by Rossetti's brother William as:

---

66 Quoted in Leonard Shengold, *If You Can't Trust Your Mother, Who Can You Trust?: Soul Murder, Psychoanalysis, and Creativity* (London: Karnac Books, 2013).

A most beautiful creature with an air between dignity and sweetness with something that exceeded modest self-respect and partook of disdainful reserve; tall, finely-formed with a lofty neck and regular yet somewhat uncommon features, greenish-blue unsparkling eyes, large perfect eyelids, brilliant complexion and a lavish heavy wealth of coppery golden hair.[67]

Others were less kind. There seems to have been something unsparkling about Lizzie Siddal altogether:

She was passive. . . . This passivity helped bring them together. She trailed slowly towards [Rossetti], a melancholy doll, set in sluggish motion by the virile, expansive gestures of the warm Latin. His roar of laughter elicited from her a wan smile, his jests provoked a faint answering shade of humour, his ardour the ghost of passion. In the same contrary fashion, he loved her because she was so little responsive. No one knew what she was thinking of or if she thought at all. She had . . . the habit of "keeping herself to herself" which deepened into an unfathomable reserve on being introduced into a clever and freakish group of artists. . . . In her mournful beauty, her natural silence, her frigid apathy, she was like a statue to be warmed into life. . . . [68]

That passivity probably had much to do with the fact that she suffered both from depression and an addiction to opium. Lizzie Siddal wrote poetry (not good), drew (not well), and died of an overdose of laudanum in 1862, leaving Rossetti, who had been neither constant as a lover nor compassionate as a husband, with a burden of lifelong guilt that one has to say he probably deserved. You have the feeling that he loved the passive face, the hair, the sad and empty eyes and all he could project onto them, but not so much the woman.

67 Quoted in Russell Ash, *Dante Gabriel Rossetti* (New York: Harry N. Abrams, 1995).

68 Quoted in Jennifer J. Lee, *op. cit.*

Perhaps his most affecting portrait of Lizzie Siddal is as *Beata Beatrix* (*c.* 1864–70; Fig. 21), created after her death and inspired by the Beatrice of that other Dante. It shows Lizzie as the loved one lost, eyes closed as if in death and much more otherworld than this, with her red hair gathered behind her like the tail of a pale comet, haloed with light. Now, in Alice, or Alexa Wilding, as his model was to rename herself, he had another Lizzie, just as beautiful, and apparently offering the same useful blankness too: "a lovely face," wrote Dunn, "beautifully moulded in every feature, full of quiescent soft mystical repose. . . . but without any variety of expression. She sat like the Sphynx. . . . "

Not all redheads are fiery. Despite her "repose," Rossetti used his new model, with her lovely face and unmissable hair, without stint. He made *The Blessed Damozel* a redhead, as if in her honor (his own poem, of 1850, had spoken of the Damozel as having "hair that lay along her back/ . . . yellow like ripe corn"). He repainted her features over Fanny Cornforth's in his *Lilith* and *Venus Verticordia*, both of 1864–8, and had an agreement with her that she would sit to him exclusively. Perhaps, in fact, it was her blankness, along with her resemblance to Lizzie, that made her so inspiring a muse. Perhaps the only role she had to fulfill was to turn up on time and be decorative. Even *La Ghirlandata*, the painting in which she stars on the cover of this book, had no "underlying significance," according to Rossetti's brother William: "I suppose he [the artist] purposed to indicate, more or less, youth, beauty and the faculty for art worthy of a celestial audience. . . . " In other words, these paintings are to show off the painter, not the model. It is we, coming to them as their audience, who look for symbolic meaning in these yards and yards of red Pre-Raphaelite curls.

Certainly there is some. It's notable that it's almost always loose or loosened red hair that is depicted, hair so luxuriant that it's almost out of control. It may symbolize female sexuality in Eve or Lilith, female passion in Sandys's *Proud Maisie*, or be the inevitable attribute of fatal beauty in the same artist's *Helen of Troy*. It's suggestive of Bohemianism, the world in which these artists lived at least at the start of their careers. Many of their paintings are mythological in subject; perhaps playing on the notion of red hair as an attribute of the supernatural. They also undoubtedly give the artists an opportunity to show off their skill in depicting these twining tresses and shining locks. And they both draw and please the eye, in which case the artists are using red hair in exactly the same way as the advertising industry of the twenty-first century. In a bit of slang favored by the Pre-Raphaelite Brotherhood, which has survived unchanged from the 1840s to the tabloid press of today, it is the coloring of the "stunner." Used like this, of course, it starts to stigmatize the very characteristic that it thrusts at our attention. There's something a little prurient about the Pre-Raphaelites. Look at their treatment of Mary Magdalene. In their hands Mary Magdalene, the most important female saint of medieval Europe and, next to the Virgin herself, possibly the most empowered female figure in medieval art, becomes either merely an excuse for painting another "stunner," or in Rossetti's *Found* (Fig. 22), a prostitute for real. In his own words:

> The picture represents a London street at dawn, with the lamps still lighted along a bridge which forms the distant background. A drover has left his cart standing in the middle of the road . . . and has run a little way after a girl who has passed him, wandering in the streets. He has just come up with her and she, recognising him, has sunk under her shame upon her knees,

against the wall of a raised churchyard in the foreground, while he stands holding her hands as he seized them, half in bewilderment and half guarding her from doing herself a hurt.

The "fallen woman" of Victorian cliché, in other words, here literally slumped to the ground.

"Magdalene," as a polite-ish bit of slang for a prostitute, had been in use since the late seventeenth century at least. What makes this painting particularly rich in allusion—so rich and complex, in fact, that for thirty years it defied the artist's attempts to finish it off, and was left as it is, uncompleted, on his death in 1882—is the fact that Rossetti's eventual model for the woman in *Found* was his mistress and housekeeper, Fanny Cornforth, she of the doubtful past. Fanny was no silent Sphynx, nor a woman who sounds likely to have had much time for disdainful reserve. As she and Rossetti grew fatter in their old age together, she called him "Rhino" and he called her "Elephant." In *Found*, however, the earthly Fanny is depicted as a pitiable being, in the last stages of consumption or perhaps of syphilis, with greenish pallor and face compressed in shame and agony. Her jaunty feathered hat has fallen back, revealing her coppery hair, by which perhaps her onetime sweetheart, the drover, is meant to have recognized her. (The sweet young thing come up from the country and ruined by the wicked city is another workhorse cliché of Victorian art and literature alike. Think of Little Em'ly in Dickens's *David Copperfield*.) The work was exhibited, even if unfinished, and found many an admirer, among them Lewis Carroll, who saw it at the Royal Academy in London in 1883 and called the face of the drover "one of the most marvelous things I have ever seen done in painting." A matter of opinion, perhaps; the painting's other great point of interest here is the way it reworks the iconography of the Colmar *Noli Me Tangere* of four centuries before. Here is the Magdalene as a Magdalene, disempowered and the lowest of the low; here it is the man who touches her against her will; here she who turns away. In Rossetti's nod to Colmar, the Garden of Gethsemane has become a single withered rose, lying in the gutter, the birds in the garden a pair of London sparrows.[69]

You get the sneaking impression that neither brains nor personality were much desired in the Pre-Raphaelite artist's model. Women with either tended to become bored and to move on, sometimes, in hope, like Lizzie Siddal, by marrying the artist. Some married and were bored still. A sixteen-year-old Ellen Terry sat for the artist G. F. Watts, a close associate of the Pre-Raphaelites in 1864, when he was forty-seven, and shortly after married him. The portrait he painted of her, *Choosing*, makes glorious play of her fair, schoolgirlish features and strawberry-blond hair, and is usually interpreted as showing Ellen's rejection of the artifice of the outside world for a life of married duty, but looking at it today, with its hit-you-over-the-head flower symbolism, it's possible to read the showy but scentless camellia that Ellen holds to her nose as life with Watts, and the tiny scented violets held close to her heart as her ambitions (like Alexa's, but much more successfully so) for a career on the stage. In 1889, at the height of Ellen's fame as an actress, John Singer Sargent would paint her in costume as Lady Macbeth, during the run of Henry Irving's production of *Macbeth* at the Lyceum Theatre in London.

---

69 Béatrice Laurent, "Hidden Iconography in *Found* by Dante Gabriel Rossetti." See http://www.victorianweb.org/painting/dgr/paintings/laurent.html.

Her extraordinary dress, in a peacock's-tail palette of green crochet and blue tinsel, sparkles with more than a thousand green iridescent beetle-wings. Willful, passionate, murderous, *and* Scottish, as she holds Duncan's crown over her own head, Lady Macbeth's ensemble is completed with two plaits of deep red hair, thick as hawsers, bound with gold, and reaching to her knees (Fig. 23).

And as an example of red hair equaling ethnicity, there is also Millais's *Martyr of the Solway* of 1871. Scotland had a particular importance for Millais. Some of his most significant works of the 1850s have Scottish settings or Scottish subjects; and it was here in 1853 that he fell in love with the auburn-haired Effie Gray, then Mrs. John Ruskin. Not that the story behind the *Martyr of the Solway* is itself an uplifting tale, nor does it have anything but the most tragic of endings. The martyr was one Margaret Wilson, an eighteen-year-old girl from Wigtown in Dumfries, who in 1685 had been sentenced to death for refusing to swear an oath accepting the Catholic James II of England, and VII of Scotland, as head of the church in Scotland. Despite the fact that a reprieve had been granted, Margaret Wilson nonetheless died as sentenced, by drowning, chained to a stake in the Solway Firth.

This too is a work with a unique backstory, which can only be told with reference to Millais's *Knight Errant*, of 1870. The redhead in this earlier work, hair blown across her naked, captive body as she is cut loose from a tree by the knight who has killed one of her attackers, has been interpreted as a victim of robbery, but rape seems just as likely (an early description speaks of the woman having been "despitefully used"). And then there is the knight's extremely prominent longsword, streaked with blood. The painting, which is otherwise rather Arthur Rackham–esque in its moonlit details and forest setting, trembles with sexual activity; and in it, in its original incarnation, the woman who now looks away from us had her face turned toward her rescuer. This was heady and subversive stuff. While a naked woman's body might be laid out for the viewers' delectation, her face was apparently better turned modestly away, in shame or private humiliation (as it would be in Lefebvre's *Mary Magdalene in the Cave*, for example). Millais's treatment was in fact held up for comparison with the "idealized" form the nude took in continental painting, and presumably the critics making this comparison must have had Salon painters such as Lefebvre in mind, rather than works such as Manet's *Déjeuner sur l'Herbe* of 1863, or his *Olympia*, of the same year. The fact that this, Millais's one full-length naked female figure is almost life-size, only added to the critic's discomfort when it was first exhibited. Her thighs are dimply, her waist a little thick; "too life-like," the critics called it, "too real."[70] It failed to sell until Millais cut away the central section and replaced it with new canvas on which the woman turns her face from us. What, then, to do with a half-length study of a nude redhead, arms bound behind her back? Create the *Martyr of the Solway* (Fig. 24) with it, of course, which, in a startling bit of artistic sleight-of-hand, is exactly what Millais did.[71] Margaret Wilson in the *Martyr* was originally the top half of the damsel being rescued by the knight. But having learned his

---

70 Anonymous review. *Art Journal*, June 1870, p. 164. Quoted in Linda Nead, "Representation, Sexuality and the Female Nude," *Art History* 6, no. 2 (June 1983) 233–6.

71 To see *The Knight Errant* as it would originally have appeared, see the cunning reconstruction created by Martin Beek on flickr and displayed by http://vadimage.wordpress.com/2010/11/08/too-life-like-the-knight-errant-1870-by-john-everett-millais.

lesson, perhaps, from the reception accorded to *The Knight Errant*, in her new incarnation Millais clothes her, in a very nineteenth-century-looking blouse and, tellingly, a plaid skirt. He also removes all possibility of rescue. There was to be no knight, no Perseus for this Andromeda. The dark waters of the Solway Firth are rushing in upon her; it is this, her fate, that she turns her eyes from, rather than her own nakedness. The viewer of the painting is left in the uncomfortable position of being forced to see what Margaret does not. But set against the dark waters as they roil in behind her and in contrast to the stormy and overcast sky is her red hair, used here like a flag, to suggest (with the tartan of her skirt) both her nationality and her defiance. This is one case where the subject's red hair and the meaning of the work are indivisible.

In fact the days of the smooth and perfect limbs of the Salon nudes were already numbered. Over the Channel the French Impressionists were also creating works in which redheads abound, in Renoir's marshmallowy nudes and in the pastels of Edgar Degas of the late 1880s in particular. There hadn't been so many redheads in art since the days of Elizabeth I. Degas's pastels of women sponging or toweling off after the bath, or combing through their long hair, have been criticized for their objectifying of their subjects, their equation of these women with so many cats washing and tending to themselves. They also have what one might view as at least a semi-exploitative subtext, in that such ablutions traditionally preceded or followed intercourse, and that the women are often naked, or almost so, and their faces are again often obscured. But Degas did produce one of the best depictions, indeed glorifications, of red hair ever, in his *La Coiffure* of 1896 (Fig. 25). Here one woman (older, a redhead, in apron and pinkish blouse) is combing through the hair of another (younger, wearing a red robe) who sits before her. The younger woman's long red hair is stretched between them, like washing going through a mangle. Everything in the image is red, from the curtain looped up in the top left-hand corner, to the color of the wall behind them. The beads of jewelry on the table are red. The young woman's cheeks are flushed. The painting shows what we should presumably take for a domestic space as being as red as a womb. It is as if the electricity one can almost hear crackling off that hair as it is combed has suffused the entire canvas with its color.

But the artist of red hair par excellence in this period must be Toulouse-Lautrec. All three of his most famous sitters from among the singers and dancers of the Folies-Bergère—Yvette Guilbert, Jane Avril, and La Goulue—were redheads. In his *Rue des Moulins* of 1894 he depicts one of his favorite models, a snub-nosed prostitute who was apparently named Rolande waiting insouciantly in line, shift held up above her navel, to be inspected for signs of syphilis. Her bright red hair is the hotspot of the painting. And of course this is another reason why artists place redheads in their works—for that vibrant dash of color, that ability of red to draw the eye; which is exactly the same reason why for a woman of the streets, or a tart in a brothel of the Belle Epoque, red hair works. It gets you noticed.[72]

---

72 The same principle has been used in the cinema, too. There is the little girl in her red coat in Steven Spielberg's *Schindler's List* (1993); and the heroine, Lola, in Tom Tykwer's *Run Lola Run* (1998), where her bright red head of hair becomes the pivot around which the film's alternative scenarios rotate.

Toulouse-Lautrec's paintings record a love affair between a particular kind of throwaway French chic and the blazingly artificial dyed red hair that can still be seen on the streets of Paris today. How did a characteristic once so linked to the lower end of the social scale become desirable and fashionable? One answer is the link forged by these artists between the image of the intriguing, independent, unconventional, Bohemian young women, and red hair either real or dyed; another can be found in a gradual change of attitudes toward red hair, evident in at least two specific examples of this period. The first is the courtesan Cora Pearl, a *grande horizontale* of the old school who was also known as "La Lune Rousse" (Fig. 26).[73] Her enormous wealth (in the 1860s she could command as much as 10,000 francs for an evening in her company) made everything she wore and every aspect of her show-stopping style worthy of emulation. Cora understood the importance of creating a spectacle in order to stay in the public eye, of reinventing her image; dyeing her hair on occasion not only red but also lemon yellow, to match the upholstery of a new carriage, and her dog blue, to match her gown. Rather more respectable was the opera singer Adelina Patti, who did much to popularize and make acceptable the use of henna as a hair dye and whose career was at its height in the 1870s and 1880s. But both women were still exploiting the ability of red hair to draw the eye and get the wearer noticed. As were Yvette Guilbert, Jane Avril, La Goulue, and Rolande her humble self. *Madame X*, the profes-

73 The name may also play on another meaning of the phrase in French, of an early spring moon before the last frosts had passed—deadly for young sprigs, just as Cora was to the young men she captivated, one of whom shot himself dead on her doorstep in despair. The character of Joséphine Karlsson in the French TV series *Spiral*—ruthless, manipulative, redhaired, and deadly—is another *Lune Rousse*.

sional beauty Virginie Amélie Avegno Gautreau, star of John Singer Sargent's portrait of 1884, was another much-emulated celebrity of fin-de-siècle Paris, also known to use henna to tint her hair. By 1881 Miss Maria R. Oakey could write of red hair in her *Beauty in Dress* not as something to be played down or even disguised but as a specific and desirable type, with its own palette of colors to show off the hair to best advantage: "White, of a creamy tone, black, invisible green [one assumes she means eau-de-nil], rich bottle-green, rich blue-green, plum color, amethyst," and so on. And by 1910, in *The History of Mr. Polly*, H. G. Wells can present an audacious red-headed schoolgirl as the object of the youthful Polly's chivalrous and most ardent admiration, and as an entity whose desirability would be wholly understood by his readers.

There is even a new vocabulary, to mark red hair's new status as socially acceptable. In 1890, on April 1 (All Fools Day) the Auburn Printing Works of Lightcliffe, Yorkshire, published *The Philosophy of Red Hair*. This work, of heavy Victorian humor, records the hapless state of mind of Rufus, a young man with red hair. Rufus reads in his sister's journal that "red hair if straight denotes ugliness," but if "given to curl" it denotes "deceit, treachery, and a willingness to sacrifice old friends for new or personal advancement." When traveling by train, Rufus is warned not to stick his head out of the window, lest he be mistaken for a danger signal. When invited to a costume party he is advised to wrap himself in brown paper and go as a lit cigar. He is presented with all the usual reasons for being an object of such ridicule: the dislike of red hair recalls the fear of the Danes; that Judas was a redhead; that it is a primitive characteristic (a favorite of the nineteenth-century anthropologist, this one). Yet at the same time Rufus notes that the typical female flirt is always presented with flaming red hair and green eyes as part of her charm. And most unfair of all: "A very curious trait with authors is that the red of the red-haired girl is transmuted to auburn, or golden, when she becomes an interesting young lady, whereas the red of the red-haired boy remains red to the end of the chapter." This is auburn in its modern meaning, an acceptable alternative to the pejorative carrots, or ginger, for a hair color that for women was well on its way to becoming positively desirable. "Auburn" has cachet.[74] "Titian," which also came into use in the late nineteenth century, according to the *Oxford English Dictionary*, has the same, plus the added advantage of suggesting familiarity with high art. Unsurprisingly, the term caught on. *The Dundee Advertiser* would note in 1904 that "twenty years ago hair with a reddish tinge was called 'carrots,' now Titian-colored locks are reckoned a definite beauty." By the time Elinor Glyn, the (green-eyed, red-haired) sensationally successful and knowingly risqué novelist of the pre–World War I period, was writing *Red Hair* in 1905, she could have her endearingly ditzy heroine Evangeline declare herself a social outlaw, "a penniless adventuress with green eyes and red hair" who is "bound to go to the devil" because of it, yet nonetheless also have her virtuous enough to not only frustrate the seductive wiles of her guardian but win over the hero's wealthy and aristocratic uncle and gain his blessing on their marriage. (And this despite the sight of her come-hither, pink silk nightdress scandalizing his killjoy relatives.)

---

74 Or, as the American humorist Mark Twain, a redhead himself, would put it in *A Connecticut Yankee in King Arthur's Court* (1889): "When red-headed people are above a certain social grade their hair is auburn." Until the 1980s, red-haired Barbie dolls were also sold as "Titian."

The red-haired flirt as a toned-down, more socially acceptable version of the red-haired Pre-Raphaelite femme fatale was a new development for a new century, and owes much to Mrs. Glyn. But even more tellingly, by the time *Red Hair* was filmed, in 1928, complete with a very early Technicolor sequence (to do justice to the heroine's red hair, perhaps?) and starring Hollywood's first redheaded sexpot, Clara Bow, Clara's fellow starlet Mabel Normand could declare in an interview, "I'm shanty Irish"—and be proud of it.

A Scotsman, Tam Blake, was perhaps the first of the Celtic diaspora to make it to the New World, in 1540 (although there are legends that the aptly named St. Brendan the Navigator beat both the Vikings and Christopher Columbus to it, in the sixth century, thus opening up a whole new field of possibilities for those Native American stories of red-haired giants). Tam would be followed by more adventurers in both the sixteenth and seventeenth centuries. Red hair—as with the Normans in Sicily—has always been a convenient marker of human migration. So it was to be again. In the 1650s, under Oliver Cromwell, tens of thousands of Irish were shipped as slaves to the West Indies. In the eighteenth century many more were transported as convicts to Australia. In what has been called the final act in the great Celtic diaspora, in the nineteenth century there were immense migrations from both Scotland and Ireland to North and South America, to Canada, and to Australia and New Zealand. Once again, the gene for red hair went with them.

Clara Bow herself, the "It" girl of Hollywood in the Roaring Twenties (a sobriquet coined for her by none other than Elinor Glyn) owed her head of bubbling red curls to her Anglo-Irish and Scottish ancestry. Bow emerged from a childhood and adolescence of great hardship and tragedy with an underlying sense of herself as being set apart and thus extremely vulnerable, which she sought to disguise for all she was worth. She spoke of herself as being awkward and funny-faced and was teased for her coppery ringlets at school as many redheaded children are, yet in front of a camera she had "It" and to spare. Hollywood, and America, had discovered the redhead.

It is one thing to have your hair color as a badge of your underclass status in the Old World. It is quite another to carry it as a marker of your identity into the New. Transpose red hair into this environment and it comes to mean something completely different. It becomes a mark of authenticity as well as of identity. Rather than a stigma, it becomes something to celebrate, a bold visual claim to your heritage and history. And it must be remembered that in America there was already an underclass, marked out by the color not of its hair but of its skin. Immigrants from the Old World might indeed be regarded as of lowly social caste by those white-skinned Americans born in the New World, just as the Irish has been by the Anglo-Irish in the Old. They might still suffer what Noel Ignatiev in his *How the Irish Became White* defines as the "hallmark of racial aggression, the reduction of all members of the oppressed group to one undifferentiated social status." But whereas in the Old World there had been no buffer, no slave community beneath them, in America there was. In fact there were two: first the black community, and

Fig. 12: Jan van Eyck, *The Crucifixion*, c. 1435–40. This painting is twinned with a right-hand panel showing the Last Judgment. It is the Crucifixion as narrative, and has been aptly called almost an eyewitness account. The Virgin is almost indistinguishable, shrouded in her blue robes as if these are her grief; Mary Magdalene, identifiable by her red hair, lifts up her hands in a gesture that combines horror, pity, and entreaty all at once.

Fig. 13: Jules Joseph Lefebvre, *Mary Magdalene in the Cave*, 1876. Now in the Hermitage, this work is typical of the highly finished Salon style of painting—so establishment on the surface, and at the same time so ripe for decoding and reinterpretation.

Fig. 16: Elizabeth I, painted by an unknown artist around 1575. Now in the National Portrait Gallery, London, this portrait epitomizes both Elizabeth's image and her elegance. The choice of colors in her dress and her accessories is perfect for a redhead, the masklike face somehow both ineffably sad and totally remote.

Fig. 17: Robert Peake's *Procession Picture of the Queen, c.* 1601. One of the last public images of Queen Elizabeth I, who was to die in 1603, showing her in one of the paler-colored wigs she favored at the end of her life. This is a snapshot of the Elizabethan court at the end of her reign, with the queen surrounded by just about all the men of influence and power of the day, bearing her aloft. At her death, an effigy, gowned and wearing one of her wigs, was carried on her coffin.

Fig. 18: James Abbott McNeill Whistler, *Symphony in White, No. 1: The White Girl,* 1862. Whistler came to object strongly to the notion that his paintings had any meaning beyond being art. He described this portrait of Joanna Hiffernan as " . . . a woman in a beautiful white cambric dress, standing against a window which filters the light through a transparent white muslin curtain—but the figure receives a strong light from the right and therefore the picture, barring the red hair, is one gorgeous mass of brilliant white."

Fig. 19: Gustave Courbet, *The Sleepers*, 1866. The naked curves, the tumbled, loosened hair… Joanna Hiffernan and an unknown dark-haired model, posed by Courbet for the delectation of Halil Bey, and for the rest of us ever since. The painting is now in the Petit Palais in Paris.

Fig. 20: Algernon Charles Swinburne, painted by G. F. Watts in 1867. At the height of his alcoholism, in 1879 Swinburne would be taken into to the Putney home of the poet and critic Theodore Watts-Dunton, who also provided a refuge for Henry Treffry Dunn after the latter quarreled with Rossetti.

Fig. 21: *Beata Beatrix* (1864–70). Lizzie Siddal as painted by Dante Gabriel Rossetti after her death. The red-robed figure might be taken as Dante Aligheri, waiting to escort her in the Underworld; the sundial is a centuries-old emblem of mortality. The poppy in the dove's beak alludes to her death by an overdose of laudanum.

Fig. 22: Dante Gabriel Rossetti, *Found, c.* 1869 (uncompleted). The calf is meant to typify the woman's plight. Rossetti published a highly charged poem, "Found," in his *Ballads and Sonnets* of 1881; the real Fanny Cornforth, who posed as his model, was photographed proudly wearing the gold earrings of the figure here.

Fig. 23: John Singer Sargent, *Ellen Terry as Lady Macbeth*, 1889. Terry wrote of the painting, "It's talked of everywhere and quarreled about as much as my way of playing the part. . . . Sargent has suggested in this picture all that I should like to convey in my acting."

Fig. 24: Sir John Everett Millais, *The Knight Errant* (1870) and *The Martyr of the Solway* (1871). The critics had a field day with *The Knight Errant*, seeing evidence on the woman's body of "the ligature of draperies," meaning she had been undressed, and in her face a character not "over pure" or "refined." In other words, whatever is supposed to have happened to her, she asked for it.

then as another wave of immigrants reached the New World from the Old, Russian Jews fleeing the pogroms of the 1880s.[75] The Irish in America could lift their social class simply by crossing the Atlantic. All the mighty difference created by their doing so is wrapped up in the one American concept of the despised "redheaded stepchild." When the Irish first began to arrive in numbers, the black community found itself referred to as "smoked Irish." The Irish in turn heard themselves stigmatized as "blacks turned inside out."[76] The Historical Society of Pennsylvania records the complaint of one black laborer: "My master is a great tyrant. He treats me as badly as if I was a common Irishman." The Irish had to decide, in effect, if they were going to identify with the slaves or with the oppressors. They chose the latter, and no difference, as Ignatiev points out, is fought for more fiercely than the thinnest. The phrase "beat you like a redheaded stepchild" in the States goes back very far and to somewhere very ugly, but I don't believe it has anything to do with the original lowly status of the Irish at all, nor as such with red hair. It indicates instead a child of mixed race, and originally very possibly the offspring of a black slave woman and a white slave owner. Hence the following exchange from Harper Lee's *To Kill A Mockingbird*, between Jem and his sister, Scout:

"Jem," I asked, "what's a mixed child?"
"Half white, half colored. You've seen 'em, Scout. You know that red kinky-headed one that delivers for the drugstore. He's half white. They're real sad."

---

75 Hollywood was to create something of a meme around this in the 1920s with films such as *The Cohens and the Kellys* (1926) and *Abie's Irish Rose* (1928).

76 See Ignatiev, *op cit*. Except of course that the word used wasn't "black."

"Sad, how come?"

"They don't belong anywhere. Colored folks won't have 'em because they're half white; white folks won't have 'em cause they're colored, so they're just in-betweens, don't belong anywhere."

In other words, unless they did something about it, the lowest and least-loved of the low. The magic of codominance means that the gene for red hair can certainly manifest itself with a black heritage—the red tint to Malcolm X's hair came from a Scottish grandfather. Nonetheless, and despite this unlovely phrase, there are, for me, very pertinent differences in the responses to red hair from one side of the Atlantic to the other. In the United States, where the legacy of black slavery has meant that social awareness of any sort of stereotyping as undesirable runs much higher than it does in Great Britain, I have often been told that there is no such thing as discrimination against redheads in the States (although many an American redhead may feel they reserve the right to disagree). Even more striking, however, are the differences in attitudes toward red hair between Canada and the States (where white immigration in the nineteenth century was elective), and Australia, where there was no urbanized underclass "other" already conveniently in existence, where immigration was often the result of judicial punishment, and where red hair carried with it all the connotations marking one out as very likely having been transported there not as an underclass but as a criminal.

Red hair in women might have become newly popular in the Old World in the nineteenth century, but there was no answering change in attitudes toward the redheaded man. In fact this period saw the creation of two of the nastiest redheaded wrong'uns ever to fall from between the pages of a book: Uriah Heep, in Charles Dickens's *David Copperfield*, of 1850, about as fell a villain as Dickens ever created; and in 1898 the character of Peter Quint, in Henry James's *The Turn of the Screw.*

Heep, as he is introduced to us, already has his ginger hair shorn convict (or we would say skinhead) short (Fig. 27). He has in his face, "in the grain of it . . . that tinge of red which is sometimes to be observed in the skins of red-haired people." In other words, he is another take on the ruddy-faced, red-haired impenitent thief of the Crucifixion. We are to understand that Uriah's true nature—grafting, scheming, leading every soul around him astray, if he can, like the devil in a medieval morality play—is announced by that telltale red coloring, seeping through the grain of his skin. Heep is sexual. He both fascinates and repels the hero, David Copperfield, in equal measure; he is Machiavellian, and he is very nearly triumphant. In 1898, Henry James would also draw upon this ancient connection between the color red and the whiff of brimstone for the character of Peter Quint.[77] Quint has a particularly sickly connection to the two young children in *The Turn of the Screw*, and he has returned in death to haunt both them and the novel's narrator, an unnamed governess. In the governess's description of him, Quint's red hair not only serves

[77] The connection between the color red and the infernal is still being used today, in the *Hell Boy* comic book and film franchise, for example.

as a marker of his diabolical nature but identifies him, even if he should be in his grave, to her listener:

> . . . he has red hair, very red, close-curling, and a pale face, long in shape, with straight good features and little, rather queer whiskers that are as red as his hair.

Heep and Quint are nineteenth-century embodiments of far older fears, and in both cases their red hair marks them out as being apart from the normal, law-abiding, or even laws-of-nature-abiding society in which they wreak such havoc. These two come from somewhere very different indeed.

But there is a change of heart toward red hair in literature for children. First in France, in 1894, there is Jules Renard's semi-autobiographical *Poil de Carotte*, or *Carrot Head*, as it is usually known in translation; and then in the New World, *Anne of Green Gables*, in 1908.

Poil de Carotte is not a children's book, even though it is a stalwart of French literature classes in school. It is instead a book about a child, and what it is to be a child growing up in a household where your father hates your mother, and your mother takes out her hatred of her own life on her youngest, redheaded son. The fact that Poil de Carotte (he is hardly ever referred to by his given name, Francois) is a redhead is almost hidden; what the book does is present the interior workings of his life, his insecurities and fears, his attempts to make sense of the world as being no different to those of any other child. He is neither more nor less disobedient, neither more nor less hot-tempered. He is a heartbreakingly human and unhappy little boy who simply happens to have red hair. Meeting the actress

Sarah Bernhardt, once he too had become a celebrity, Jules Renard (fittingly, his surname means "fox") recorded in his *Journal* a faltering recognition of the shift to a more sympathetic awareness to which his book had contributed, when the divine Sarah tries to excuse the fact that he is a redhead: "Redheads are ill-natured. . . . You are rather on the blond side."

And then of course there is Anne. As envisaged by her creator, Lucy Maud Montgomery (a Canadian—would Anne have been the same had she been conceived in the Old World?), "she wore a faded brown sailor hat and beneath the hat, extending down her back, were two braids of very thick, decidedly red hair. Her face was small, white and thin, also much freckled."[78]

Anne, like David Copperfield, is an orphan, and as a character is related to those formulaic orphan children of so many children's stories (get the troublesome parents out of the way and have the hero or heroine stand in their own light), but Anne has a brain and a temper. Long before the ineffable Australian comic Tim Minchin, and his song, "Only a Ginger Can Call Another Ginger Ginger," Anne hits the nail exactly on the (red)head: "There's such a difference between saying a thing yourself and hearing other people say it." When Anne's coloring is commented upon in her hearing by the unpleasant Rachel Lynde ("Did anyone ever see such freckles! And hair as red as carrots!"), she fires up in her own defense at once.[79]

---

78 Anne was supposedly based on Evelyn Nesbit, a famous pin-up girl of the period, another onetime artists' model, and the center, in 1906, of a world-famous scandal (dubbed the "Trial of the Century") when her husband shot dead her lover, the renowned architect Stanford White, in Madison Square Garden. Notwithstanding, she was Montgomery's inspiration.

79 A perfect example of why prejudice against red hair is so pernicious. Thanks to this business of there

Yet, for all her spirit, Anne too longs for her hair to be "a handsome auburn" when she grows up.

Anne's quick wit and precocity are, if you like, an acceptably infantilized version of the red-haired flirt (Evangeline, Clara Bow), the girl with a twinkle in her eye and a smart riposte but with her virtue intact. Via Anne, the redhead female juvenile lead descends to Little Orphan Annie herself, in the 1920s, and to the redheaded tomboy's all-time heroine Pippi Longstocking, living adult-free in her Villa Villekulla with her monkey and her horse. Pippi was first published in 1945; to my mind Jessie the Cowgirl of *Toy Story 2* (1999) owes much to Pippi.[80] Thence and more recently we come to the independent-minded and very freckly Freckleface Strawberry of today. And let's not forget the Little Red-Haired Girl in *Peanuts*, object of as devoted an adoration on the part of Charlie Brown as ever Alfred Polly suffered for his red-haired sweetheart, sitting on her school wall, forever out of reach.

It's children's fiction that also finally creates a male redheaded character neither evil, ill-intentioned, nor a milksop. Tintin, first published in 1929, is resourceful, adventurous, and intrepid and has a signature cowlick of red hair (Fig. 28). And, he is the hero. William, of the *Just William* stories of the 1920s, has his trusty sidekick, Ginger, but he is a sidekick only. (William also has a redheaded and

---

being no "cultural barriers," no obvious visual difference between the victimizer and the victim, just as there are none between Rachel Lynde and Anne, it doesn't look like the prejudice it is.

80 Following criticism of Barbie's sexually mature face and figure, early versions of Midge were closer to the tomboy model, with freckles and a rounder face, in an attempt to make her (and Barbie) seem less adult. Midge today is rather more of a flame-haired siren than she was. The same freckles equals cute and wholesome theme is at work in the logo for the Wendy's fast-food chain.

at least by implication devastatingly soigneé and attractive older sister, Ethel, whose popularity with any number of eligible bachelors in the neighborhood is an eternal mystery to her younger brother.) *Biggles*, first published in 1932, also has a "Ginger" as his sidekick, thus continuing a tradition still used by writers today—think Ron Weasley, second-in-command to Harry Potter. But Tintin, despite being the youngest character in the books, is the protagonist, has a personality full of all the stalwart virtues of the Boy Scout, and is a first. Do these juvenile leads offset the centuries of prejudice against redheaded men in art and literature, their use as shorthand for villainy of every sort? I think that they at least start such a transformation, and they certainly record a shift in society's attitudes, away from "redheaded woman good/redheaded man bad" to something less unthinking and rather more nuanced. But it takes many, many Tin-Tins, Gingers, and Rons to expunge the centuries-long prejudice reflected in a single Uriah Heep. And what's Heep's come-uppance, when all his schemes and machinations are foiled? Transportation for life, of course—to Australia.

RAPUNZEL, RAPUNZEL

Well, listen up, stud,
Your life's been wasted
'til you've got down on your knees and tasted
A red-headed woman . . .

BRUCE SPRINGSTEEN, "REDHEADED WOMAN"

Augustin Galopin was a French man of letters, of philosophy, and of medicine. By 1886 he had published more than twenty works on subjects ranging from cremation to feminine hygiene. The health and well-being of the female sex was clearly a matter of great concern to Dr. Galopin, who from his writing it is very easy to imagine as a dapper Fernando Rey type, strolling along through the Luxembourg Gardens, savoring the smell of the horse-chestnut flowers (which, according to the Marquis de Sade, smell of spunk) and tipping his hat to any particularly *jolie dame* who happened to catch his eye. This was the Paris of Toulouse-Lautrec, after all. It was also the city in which Dr. Galopin let loose his *Le Parfum de la Femme* upon an unsuspecting world.

Le Parfum de la Femme is a winning mix of folk wisdom and high-blown science, spiced with anecdotes and Dr. Galopin's own observations and musings. He informs us for example as gospel truth that a plant, the Dutchman's pipe (*Aristolochia*), has often been observed to kill serpents—*Aristolochia* is remarkably toxic and possibly carcinogenic, but its anti-serpent properties are unproven, to say the least—and he then notes as an aside what sounds like a proper empirical scientific observation: that if you have been handling female toads, and place your hands in water, male toads will rush to them. Galopin claims that he is a materialist, hence all is to be proved by experiment and observation, which makes some of his assertions more than a little hard to take. While assuring us that exposure to tobacco provokes St. Vitus Dance, or chorea, and masturbation in children, he also claims to have broken the addiction to tobacco by substituting coffee and sugar (which for the anxious parent hardly sounds like an improvement). And then there are his views on female odor.

Galopin believed, rather charmingly, that each woman, classified by skin and hair color, had her terroir, like a wine, and gave forth a specific bouquet of scents. For example, women with chestnut hair, so he assures us (those chestnuts again!), give forth an odor of amber. Perfumiers do recognize a type of scent classified as amber. It's chiefly composed of vanillin and labdanum, a plant resin.[81] The good doctor may have meant this, as vanillin was a creation of the late nineteenth century, but as Galopin uses the term, and in the context of the passage paraphrased above, it seems more likely that he was talking about "amber" as shorthand for "amber*gris*."

Now ambergris, for those who have not sampled it, has an odor

---

81 Thanks to http://perfumeshrine.blogspot.com for this information, and many another fascinating insights into the history of scent.

so strong it's as if the human nose doesn't know what to do with it. It is salty and marinelike with base notes that are, frankly, fecal. To put it bluntly, if the sea emptied its bowels, the result would smell like fresh ambergris, and unsurprisingly, too, when you consider that this extraordinary substance is produced as some mysterious digestive aid in the stomach of sperm whales. Once it has aged and oxidized, however (which can take years—ambergris is soft and waxy and it floats, and ideally, once voided by a whale, would spend those years bobbing about in the ocean), its smell sweetens, although it never quite loses that animal tang of the midden, to something subtler and, for want of a better word, hormonal. It is impossible to smell ambergris and not think of sex. Its special value to the world is that it is an excellent fixative and carrier of other scents, making them last much longer, hence in an age when the science of artificial perfumes was in its infancy, it would have been much more commonly encountered than it is today. But according to Dr. Galopin, this was the natural scent of women with chestnut hair. "And some women with that hair color, who have very white skin, exhale a soft odor of violets from most of their sebaceous glands." Then "when they are hot . . . the coquettes pretend they don't know the ravages their perfume molecules make in the brains of those who breathe them in." The little minxes, shame on them. And that, you might imagine, would be the end of the matter; simply another pseudoscientific myth about red hair. You would be wrong. Redheads *do* smell different. Or rather, if you have red in your hair, anything applied to your skin is going to smell different from the way it will smell on anyone else.

The biochemistry of the human animal, as modern science is starting to unravel its secrets, is more complex and more fascinating than anything even Dr. Galopin might have imagined, and to those in the field it must sometimes feel as if every discovery yields up yet another mystery. Biochemists and geneticists are in something akin to the situation of explorers, attempting to understand the layout of a lost city in a jungle. Each starts in a different place. Each has a piece of a map. Gradually the main roads and byways and connections both expected and unimagined begin to emerge. And in the case of the boulevardier-biology of Dr. Galopin, the most surprising claims, the oldest of old wives' tales, can prove to be entirely correct.

Much of this, particularly for us layfolk, lies at a level of microscopically complex science that can best be dealt with by employing the principle of "Don't Worry About It."[82] We do not need to be capable of understanding, at a glance, the information that MC1R is a seven-pass G-coupled receptor located at chromosome 16q24.3 (although it is pleasing to have what you might call our red hair's postal address), or indeed that it is part of family of genes, from MC1R to MC5R. What is of interest here is the list of biological functions along with hair pigmentation in which these genes play a part. Among them are adrenal function; responses to stress; the fear/flight response; the pain and immune response; energy homeostasis (the body's chemical regulation of its use of energy); and sexual function and motivation. Briefly, all these fundamental functions of the human body are different for redheads from those of blonds or brunettes as a result of redheads' uniquely different biochemistry,

---

82 Professor Sydney Brenner, who won the Nobel Prize for medicine in 2002, created this hypothesis in 2011. In his words, "It forces you to keep going without losing your mind over mechanisms." *Metode*, interview with Sydney Brenner, http://metode.cat/en/Issues/Interview/Entrevista-a-Sydney-Brenner.

the consequences of which have all fed into the stereotyping and societal and cultural attitudes evinced toward the redhead for centuries. And while these differences are far more than skin-deep, that is where we start.

We all, on the surface of our skin, have a microscopically fine film known as the skin mantle. It acts as a barrier to bacteria and other contaminants, and on those possessing the gene for red hair it will very often be more acidic than on the skin of a blond or brunette. This is why any scent or cologne will smell different on a redhead from the way it smells on his or her non-redheaded girl- or boyfriend.[83] Nor does scent last as long on a redhead's skin as it does on that of a blonde or brunette, causing Dr. Galopin to bewail the fact that in the heady days of the first artificial perfumes, redheads used so much synthetic scent and products so concentrated "that they asphyxiate all those who approach them." The havoc redhead skin wreaks on the products of the perfumier's art is strange enough, but it also suggests another reason, one with a flawlessly scientific basis, for the perceived sensuality of the female redhead: pheromones. Those same sebaceous glands whose "perfume" molecules so disordered Dr. Galopin, and which secrete the skin's acid mantle, also produce pheromones, most generously from those parts of the body that still retain our original covering of hair, that is, the genitals and under the arms.

Pheromones—messages in a smell, basically—are an invisible

---

83 Susan Irvine, *The Perfume Guide* (London: Haldane Mason, Ltd., 2000), 9. Opium, on me, for example, the notorious spice-bomb scent of the 1980s, smells like a cat that has just come back from the vet's. If you want a scent created for redheads, and can find it, the Perfume Shrine blog recommends Patou's "Adieu Sagesse," created in 1925. It has a charming story and celebrates the throwing of caution to the winds in a love affair.

Morse code by which we share signals about our general state of health, and our receptiveness to a mate, unwittingly and with the whole of the world around us, twenty-four-seven.[84] If, on a redhead's skin, the scent from a bottle can be so transformed by the chemicals produced by their sebaceous glands, I suppose it's possible that redhead pheromones also differ, uniquely, from those of blondes or brunettes, and send forth some secret message of their own. There are endless articles on endless websites dedicated to promulgating the idea that redheads have more sex than blondes or brunettes, that sex with a redhead (inevitably a female redhead) is somehow "better" or more passionate, and that part of this derives from a redhead's particular bouquet. It has become a literary device. Jean-Baptiste Grenouille, the antihero of Patrick Süskind's novel *Perfume* (1985), commits murder to capture the scent of a redhead:

> Her sweat smelled as fresh as the sea-breeze, the tallow of her hair as sweet as nut oil, her genitals were as fragrant as the bouquet of water-lilies, her skin as apricot blossoms. . . . The harmony of all these components yielded a perfume so rich, so balanced, so magical, that every perfume Grenouille had smelled until now . . . seemed at once to be utterly meaningless.

Aristide Bruant, a cabaret singer and friend of Toulouse-Lautrec (he is the man in the red scarf in the artist's famous *Ambassadeurs* poster) contributes this, from his 1889 song "Nini Peau d'Chien":

She has soft skin
And freckles

---

84 Many years ago I was propositioned by a colleague who was married to one redhead, in the midst of a torrid affair with another, and declared he had reserved a room at his club for the two of us. Even for me, this was redhead overload. Refusing his offer as graciously as I could, but nonetheless intrigued, I asked him what was this thing with him and redheads. He answered, somewhat sheepishly, "You smell different."

And the scent of a redhead
That gives you the shivers.

Before either of them there was Charles Baudelaire and his beggar-girl, the "Blanche fille aux cheveux roux," whose body, dotted with freckles, "has its sweetness."[85] (The beggar-girl of the poem, with her dark chestnut locks, was painted by Baudelaire's friend the artist Émile Deroy around 1843–5. You can see her portrait in the Louvre.) And if we smell different, might we taste different as well? I have no idea, but The Boss seems to think so— and he would be a man to know.

We need to talk about sex.

The psychiatrist Charles Berg was a Freudian of the old school. To give you a flavor of his writing, from *The Unconscious Significance of Hair* (1951): "The Christmas tree has been associated with hair and with father's penis. We see Father Christmas (with his long beard) taking off the tree penises, which he benevolently gives to children. At the feast, the phylogenetic successor of the old totem feasts, his penis is eaten in the shape of an appropriate symbol— turkey or goose." (I promise I am not making this up.) Part of *The Unconscious Significance of Hair* deals with case studies, one of which,

---

85 From his collection of poems *Les Fleurs du Mal* (1857). Another poem from the same collection, "Delphine et Hippolyte," was a part of the inspiration for Courbet's *The Sleepers*.

as related by Dr. Berg, includes a dream told to him by a young male patient, in whose psyche Dr. Berg was rooting for any number of unacknowledged neuroses. In this dream, the young man was sitting on a bus, and putting out his hand to touch the red hair of the woman sitting in front of him, experienced what Dr. Berg describes as intense pleasure, and which you or I, if we wanted to be equally euphemistic, might term a visit from Lady Lilith. In other words, an erection, if not an ejaculation. And why? Because the young man had recently succeeded in convincing his girlfriend to let him remove her underwear (this was the 1950s, let's not forget) and on first seeing her pubic hair, had been delighted to discover that it had a reddish tint. Dr. Berg, I suspect, was not a fan of the redhead (according to him, redheads had "a supernormal capacity for rheumatism, chorea and TB," as well as "detumescence") and he notes nothing significant, unconscious or otherwise, in the young man's reaction. You or I might beg to differ. In fact I'll hazard a guess that the reason for the young man's delight was that, however respectable his girlfriend might be on the surface, this was proof-positive that in secret, and known only to him, she was hot stuff.

What *does* red hair mean? Not what does it mean for a redhead, but what does it mean for everyone else? Above all, what is this mysterious connection made with such constancy between redheaded women and sex? Or, to quote the writer Tom Robbins, in a favorite paean of redhead prose, "How are we to explain the power these daughters of ancient Henna have over us bemused sons of Eros?"[86]

---

86 The entirety can be found at http://www.angelfire.com/az/varuna/ode.html.

We've already speculated that red hair could have indicated to a mate that you would bear healthy children, and not die yourself in the process. Dr. John Cook, writing for the *Ladies' Magazine* in June 1775, offers this: "Red hair is not so agreeable, though this I can say, such women have the finest skins, with azure veins, and generally become the best breeders of the nation." Let's get down to the nitty-gritty here: what does it mean when you're naked?

To begin with, phaeomelanin, the chemical that colors red hair red, also colors those parts of the body chosen by Nature to stand out as pink. That is, the nipples, in women the labia, and in men the glans of the penis. Set against a redhead's normally pale skin, a naked redhead, male or female, is thus flashing a set of sexual super-stimuli at their partner, doubly so when aroused, when the coloring in these parts of the body deepens. (I am also told that again, due to the pale skin, when a redhead reaches orgasm, the skin flush is particularly noticeable and gratifying.) Could this be another reason why, to quote Grant McCracken, red hair in a woman is seen by (male) society as promising "sensual delights of extraordinary proportions"? Redheads are rule-breakers, rebels. Grant McCracken again: "We cannot rely on them [female redheads] to embrace stereo-typed qualities of femaleness—sweetness, docility and politeness. . . . We imagine them ready to give vent to what we keep harnessed." This notion has been around for centuries. Jonathan Swift, in his *Gulliver's Travels* of 1726, has the red-haired members of his imagi-nary race the Yahoos being "more libidinous and mischievous than the rest." Red hair is a warning flag: here comes trouble. Sexually, this is a charged and potent mix. Redheads are different, so redheaded women, perhaps the thinking goes, *might* do things other women won't. . . .

There is of course a difference between the perceived and the actual experience of being a redhead, which one might liken to the difference between pheno- and genotype—except that where sex is concerned the two are not so easy to separate.[87] Does a partner's expectation affect sexual behavior? I would say yes, most definitely. If cultural attitudes toward your hair color give you license, as it were, to be articulate and confident in bed, will better sex result? Without a doubt. Will you be happier and more confident in your own sexuality, if you anticipate eliciting a positive response from your partners? What do you think?[88]

Then there is the part played by pheromones. There are good reasons to believe that we have retained body hair under our arms and around the genitals because the hair helps disperse pheromones into the air, and in the case of a redhead, one of the messages those pheromones is carrying is health. Again, this is a highly desirable quality in a mate, but, sadly, it has nothing to do with the red hair—red hair is simply the signifier; it's the pale skin that makes the difference, and the ability to synthesize vitamin D.

Most of the vitamin D we need is made in the skin, in response to exposure to UV radiation. The farther north you go, however, the fewer days there are in the year when vitamin D can be produced, and the more when vitamin D will be broken down by the body for

87 Briefly, your genotype is your genetic coding; your phenotype is a composite of that plus every other thing that acts on you to alter your observable characteristics.

88 A much-cited survey, conducted in 2006 by a Hamburg sex researcher, Dr. Werner Habermehl, appeared to suggest that redheaded women, at least in Germany, had a more active sex life than did those of other hair colors. Dr. Habermehl, however, confuses quantity with quality, which is a pretty startling error; for other issues with this survey see http://www.drpetra.co.uk/blog/do-redheads-really-have-more-sex.

use. And without enough vitamin D, everything eventually stops working. Hence the wretched fate of the Viking settlers at Herjolfsnes in Greenland.

Herjolfsnes (now Ikigait) was named for one Herjolf Baardsen, a follower of Erik the Red, and founded in 985. It was a tough and inhospitable place to live, but the Vikings were nothing if not tough themselves, and the point where the settlement was founded was the first landing place for ships that had trekked from Iceland or even Norway, so the community here should have been set to thrive. It did not. It simply disappeared. For many years the native population of Inuits were blamed for the settlement's demise. Then in 1921 the remains of Herjolfsnes were examined by archaeologists, who found there the ruins of a church and other buildings, and numerous burials, preserved by the cold.[89] Herjolfsnes is a Tarim of the snows.

The bodies in the graves told a sad story of decline. Shrouds were patched and reused; coffins, too. Those laid to rest were of noticeably short stature. The account of the excavation records with sadness how many of them were very young and notes too that "a conspicuously large number of the women were of slight and feeble build: they were narrow across the shoulders, narrow-chested, and in part narrow at the hips. Several showed symptoms of rachitis [rickets], deformity of the pelvis, scoliosis and great difference in the strength and size of the left and right lower extremities." Their teeth were worn from eating hard vegetable matter—these were people who had existed on a starvation diet for generations. It's been

89 William Hovgaard "The Norsemen in Greenland: Recent Discoveries at Herjolfsnes," *Geographical Review*, 15, no. 4 (October 1925): 605–16.

conjectured that by the time they needed to, the Viking settlers were too enfeebled to hunt seal and fish as did the Inuit. The climate had worsened. Fewer children were born. Those who were died young. The longboats from Norway had ceased. The community died of slow physical deterioration. It was killed, in fact, by the long Arctic winters and a lack of vitamin D, from which the Inuits' meat-rich diet protected them. The body of the last Viking was reportedly found by the Inuit in 1540, on the floor of his dwelling, where he had died, alone, his sheath-knife "much worn and wasted" by his side.

This is salutary enough. But vitamin D deficiency has also been implicated in an increased risk of certain cancers, of hypertension, cardiovascular disease, diabetes, autoimmune diseases such as multiple sclerosis, rheumatoid arthritis (*pace*, Dr. Berg), irritable bowel syndrome, and gum disease. Its role in preventing rickets has been known since the 1930s. And in the days before we all lived long enough to be so troubled by cancers of one sort or another, it was a weapon in the arsenal against the then all-time killer, tuberculosis. Plenty of vitamin D gives you a stronger immune system all around. So if your red hair comes with that pale skin, one of the messages your pheromones will certainly be carrying will be a message of health, and of an immune system boosted with resilience.[90]

Pale skin has also, in both East and West, been prized for centuries as an attribute of female beauty. It speaks of seclusion, of being kept apart, of Rapunzel's tower, and, to a degree, of ownership. It meant you did not have to earn your bread in manual labor.

90 A. W. C. Yuen and N. G. Jablonski, "Vitamin D in the Evolution of Human Skin Colour," *Medical Hypotheses*, 74, no. 1 (January 2010): 39–44.

It speaks, basically, of the harem, the seraglio. But in men, the message carried by pale skin is completely different, and it seems at least possible that one of the reasons why red hair is so gendered, why what is regarded as an acceptable if not indeed a highly desirable characteristic in one sex is seen as being so much less desirable in the other is because redheaded, pale-skinned men are presenting what many cultures have regarded for centuries as an attribute of female beauty.[91] Pale skin in man is a quality ascribed to the milksop (just look at the word)—someone too unhardy to go out into the world and make his way with his fellow men. In fact a 2006 study found that a significantly higher proportion than you would expect of CEOs had red hair, extrapolating from its rarity in the population at large, which may say much of interest concerning the effect on character of having to overcome being teased or bullied because of your hair color early on in life.[92] In our idiosyncratic human way, however, does the fact that something is nonsense stop it from being believed? No, it does not.

But then so much about hair is gendered, and is completely opposite from male to female. "If a man have long hair it is shame unto him." So wrote St. Paul in his first letter to the Corinthians. "But if a woman have long hair, it is a glory unto her." In his 1987 study *Shame and Glory*, the sociologist Anthony Synnott lists the following examples of contradictions between the sexes: hair on

91 Intriguing in this context that redheaded men are reportedly much more likely than women to describe themselves with the mildly pejorative term "ginger." Redheaded women are much more likely to describe themselves as "strawberry blonde." See Eleanor Anderson, *op. cit.*

92 Margaret B. Takeda, Marilyn B. Helmo, and Natasha Romanova, "Hair Colour Stereotyping and CEO Selection in the UK," *Journal of Human Behaviour in the Social Environment*, 13 (July 2006): 85-99.

the body is seen as good on men, but bad on women because it is regarded as a "male" characteristic, and is therefore with much labor and some pain removed (could this also be something to do with wanting to show off that pale, feminine skin?).[93] Men uncover their heads in a holy place; women cover theirs. Men rarely change their hairstyle—a man can have basically the same cut as he had as a boy in short trousers, which may also be pretty much the same hairstyle as his father had before him, and this is seen as entirely normal and acceptable; whereas women change their hairstyle with far greater frequency (Synnott gives a particularly telling example: when was the last time you saw a man change his hairstyle simply to go out to dinner?), and this is seen as entirely normal and acceptable for them. The norms of male hair tend toward unchanging uniformity—Synnott cites the example of a man with the same haircut JFK sported fifty years ago being wholly unremarkable today, whereas a woman styling her hair like Jackie O. would be regarded as making a statement that was very consciously retro. Women want their hair to be individual, to stand out. The conspicuousness of red hair, for a woman, its very rarity, thus makes it an advantage to her (which is no doubt why red hair dyes have such a large share of the market). It does exactly the opposite for a man, for exactly the same reason and by exactly the same means. Is this another factor behind red hair being regarded as less desirable both in and very often by the men who have it? Because it denies them that ability to

---

93 Another anomaly of being a redhead: laser hair removal rarely works, if at all, on red hair on pale skin—the laser needs the dark color of eumelanin at the root of the hair in order to heat it and destroy it. Redheads are doomed to the lengthy and unpleasant process of electrolysis if they want permanent hair removal.

blend in? Finally—and never mind numbers of redheaded CEOs—Synnott quotes another survey, this time from the United States and conducted in 1979. The finding here was that redheaded women were regarded as the executive type: brainy but no-nonsense, and slightly scary to the opposite sex (think Agent Scully: *The X-Files* played with this trope series after series); while redheaded men were regarded as "good but effeminate—timid and weak." It's as if the sexes had swapped their usual stereotypes entirely.

And there may be another reason why redheaded men have a reputation for a certain wimpishness. Once again it is a result of the redhead's unique biochemistry. Redheads feel more pain than do blonds or brunettes. Or rather, we feel the same amount of pain much more acutely, and thus require much more anesthesia to knock us out—20 percent more being the rule of thumb among anesthetists and surgeons I have spoken to. There is as you might imagine much discussion as to why this should be, how much it varies from redhead to redhead, whether some forms of pain (thermal, for example) are better or worse tolerated by redheads, and which and what anesthetic drugs are thus contraindicated. We do not bleed more than those of other hair colors. That is a myth (although I have also heard it stated as fact by one surgeon, at least, that we do). We do not bruise more easily than those of other hair colors. That is another myth, and very likely a result of the fact that bruises show up so much more noticeably on pale skin. But we do indeed require more anesthetic, and we all have horror stories of trips to the dentists as children when we weren't given enough. Unsurprisingly, redheads are notoriously bad at keeping dental appointments, having injections, and as children, having knots dragged out of our red hair (think of the poor woman in Degas's *La Coiffure*, wincingly holding on to her roots as her hair is combed out); but paradoxically, a normal level of pain for us would reduce many a blond or brunette to tears. The stoicism thus engendered nonetheless seems like a rather poor evolutionary trade-off to me. Why on earth should this pointless and painful adjunct of having red hair exist? Redheads also react badly to cold, reporting physical pain at temperatures perfectly bearable by non-redheads, although, paradoxically, we can eat highly spiced "hot" food with no discomfort at all. Madras curries and Scotch bonnet peppers hold no terrors for the redhead. There is also a belief, repeated on websites without number, that redheads are unusually susceptible to industrial deafness. Red hair is linked to brittle cornea syndrome, and there is an adrenal malfunction, one of the indicators of which is red hair, that is linked to early-onset obesity.[94]

With redheads, even the hair itself can be troublesome. Hair is made of keratin, the same substance as fingernails, and its character as well as its color derives from the shape of the follicle and the pigment-producing cells in the follicles. Beyond that, lively arguments are still under way as to why we have hair, why we have it where we do and not where we don't, and why there should be so many different types. Redheads have around 90,000 hairs on average, fewer than blonds and brunettes. Princess Merida's computer-rendered 1,500 separate strands, which would equate to about 112,000 actual hairs, are therefore well above average. Her

---

94 See http://www.ncbi.nlm.nih.gov/pubmed/9620771.

curls, however, would have to be natural; the keratin of red hair also contains more sulphur (up to twice as much) than hair of other colors, which makes it more difficult to perm. There are more disulphide linkages in red hair, which have to be broken down for the perm to take. This bolshy tendency of red hair to fight back against the hairdresser's art and its wearer's wishes has been known since the eighteenth century at least. In his *Art of the Wigmaker* of 1767, Monsieur François de Garsault explains, "Another kind of hair coming from Switzerland and England is also sold, this is red hair which has been bleached in the fields as cloth is bleached, and which for this reason is called 'field hair.' It will not frizz [that is, take a tight curl], and is only used for graduating the straight, smooth hair. It should *never*," he warns, "be mixed with a mass of frizzed hair." If you want red hair to behave, you need to show it a firm hand. Oh, and because of its color, redheads are also much more likely to be stung by bees.[95]

There is much in the strange and unusual connections afoot in a redhead's internal chemistry that in the words of Jonathan Rees still needs to be bottomed out.[96] To begin with, he notes an unexpected degree of diversity within that MC1R gene, seesawing between fully homozygous changes, where the recessive gene resulting in red hair is present on both chromosomes (a full-on redhead, as it were), and where it is only present on one and is only partially expressed (resulting in codominance and brown hair and freckles

95 Thanks to Tim Wentel for this and many another point of information in this chapter.

96 Jonathan Rees and Thomas Ha, "Red Hair—A Desirable Mutation?," *Journal of Cosmetic Dermatology*, 1, no. 2 (July 2002): 62–65.

or brown hair/red beard). Tim Wentel, a dermatologist at Erasmus University Medical Center in Rotterdam, suggests there may be as many as four hundred different genetic possibilities. Nor is MC1R on chromosome 16 any longer the only note in the chord. There is the part also played by HCL2, on chromosome 4.[97] One might anticipate that where there are two, there will be more, and as proof of the almost limitless discoveries waiting to be made as our DNA is finally uncoiled from its helix, labeled and laid out straight, you could hardly do better than point to the people of Melanesia.

The first definitions of what it was to be Melanesian were suggested by eighteenth-century European explorers of the region; even now there is no agreement on where the boundaries of Melanesia should be traced, or even whether the term is a geographic or a cultural entity. But to those people living on these islands, the word "Melanesian," which was once redolent of subjection and denigration, has become a term of affirmation and empowerment (redheads, take note). And on the Solomon Islands, 5 to 10 percent of the population, along with having very dark skin, have afros of the most striking shades of anything from a cinnamony-ginger to peroxide yellow (Fig. 29). Any number of explanations had been offered for this: natural bleaching by sun and salt water, diet, or the genetic legacy of early European explorers. But in 2012 it was found to be the result of another unique recessive gene, one totally separate from MC1R, and found nowhere else in the world. The geneticist Sean Myles, who finally identified the gene, has called it

---

97 Mutations on chromosomes 16 and 4 lead to brittle cornea syndrome 1 and 2, respectively. See http://www.omim.org/entry/229200.

"a great example of convergent evolution, where the same outcome is brought about by completely different means."[98] Sherlock Holmes and Jabez Wilson would have known all about that.

Unhappily, not all such recent genetic discoveries are quite so cheering. As with our prehistory, much of what follows is speculative and contentious. One only has to look at the cautious language of the science to realize that. Nonetheless, being a redhead can have very undesirable side effects indeed.

The biggest villain in the cast here is melanoma. Melanoma, as we all should know by now, is a particularly aggressive form of skin cancer, with a vindictive propensity to spread from the skin to other organs. One possible scientific explanation for this, for melanoma and a number of other conditions, suggests that despite the stronger immune system, redhead DNA is more fragile than that of other hair colors, less good at repairing itself, and therefore more prone to those disorders that, like melanoma, arise in damaged cells. As evidence of this, there is a link, although no one can state categorically where in the triangle is the cause and where the result, between red hair, melanoma and two serious medical conditions, the first of which is Parkinson's disease, and the second is endometriosis. Melanoma is a point in the triangle for both of them.

Parkinson's is a degenerative disease of the central nervous system. Its causes may be genetic; its development is associated with head injury and exposure to certain pesticides. In this particular

98 See http://news.sciencemag.org/2012/05/origin-blond-afros-melanesia.

Venn diagram of nastiness, Parkinson's is in one circle, melanoma in the second, and red hair in the third.[99] A history of melanoma is "associated" with an increased risk of developing Parkinson's. MC1R gene variants are associated with an increased risk of melanoma. But the lighter your hair color, equally the greater your risk of developing Parkinson's, with those with red hair being at the greatest risk of all—three times the risk of those whose hair is darkest. This is all deeply depressing stuff, no matter that Parkinson's is still a rare disease. It was thought at one time that the incidence of melanoma might be a side effect of one of the commonest treatment for Parkinson's, with the drug form of L-dopa, but the finger now points to the likely villain being the MC1R Arg151Cys allele. It probably makes this no easier to comprehend if I remind you here of those four hundred estimated possible variants of MC1R. Don't worry about it.

An equally unholy trinity exists between the incidence of red hair, melanoma, and endometriosis, another disorder of the immune system and a cripplingly painful condition where cells similar to those that line the womb (the endometrium) begin to grow on other organs in the abdomen, and just as if they were still within the womb, react to the menstrual cycle by swelling and bleeding. A study in 2000 of 3,940 college alumnae in the States found that among the group as a whole, 6.98 percent reported some degree of endometriosis. Of the 121 redheads in the group, however, this went

99 X. Gao, et al, "Genetic Determinants of Hair Color and Parkinson's Disease Risk," *Annals of Neurology*, 65, no. 1 (January 2009): 76–82; also C. Kennedy , et al., "Melanocortin 1 Receptor (MC1R) Gene Variants Are Associated with an Increased Risk for Cutaneous Melanoma Which Is Largely Independent of Skin Type and Hair Color," *Journal of Investigative Dermatology*, 117, no. 2 (August 2001): 294–300.

up to 12.4 percent, and again correlated with an increased incidence of melanoma. Or, as the researchers put it, "Among women with red hair there is an association between endometriosis and melanoma . . . which warrants further investigation." (I'll say.) They also speculate that part of this linkage may be bound up in the fact that the HCL2 gene is on chromosome 4 and thus close to the cluster of genes for fibrinogen, a protein necessary for the formation of blood clots.[100] As if redheads needed more reasons to stay out of the sun.

But the causal link, between red hair and blood and menstrual bleeding, harks back not only to the age-old slur that to have red hair means you were conceived while your mother was menstruating but to ancient Ayurvedic medicine as well, which also links red hair to disorders of the womb. Ayurvedic medicine has been around for roughly 3,000 years and is clearly akin to the Greek system of classifying humanity into types according to the four humors, which for most of recorded history was the basis of all Western medicine, too. The Greeks recognized four physiological types: you could be sanguine, choleric, melancholic, or phlegmatic, or any combination thereof, each of them being governed by one of the four "humors" or fluids within the body—respectively blood, yellow bile, black bile, or phlegm. (The origins of this system may go back even further, to Ancient Egypt or possibly to the Mesopotamians.) Ayurvedic medicine differs from the humors in that in has just three types, or *doshas*, with redheads being generally classed as the *pita* type, and being characterized as passionate, chivalrous, sensitive, and

---

100 G. Wyshak and R. E. Frisch, "Red Hair Color, Melanoma and Endometriosis: Suggestive Associations," *International Journal of Dermatology*, 39, no. 10 (October 2000): 798.

compassionate—all to the good. The *pita dosha* in women controls the health and well-being of the womb, and imbalances within it will manifest themselves there. Here is contemporary Western science confirming what traditional non-Western medicine has believed for centuries. Again according to Ayurvedic medicine, redheads are also prone to frustration, anger, arrogance, and impatience, and when out of sorts, it comes out in their skin, in eczema and dermatitis. Redheads sometimes have what is known as an "atopic" constitution (I certainly have), where the red hair goes along with a propensity for hay fever and any number of other annoying allergies as well.

One of the oldest stereotypes of the redhead is that they have a fiery temper. Again, how to distinguish between the geno- and the phenotype: if you are teased as a child you may very well react by losing your temper. If, like me, you work out that you have more leeway in displays of tantrum than your non-redheaded peers, that behavior is reinforced. But it is now thought that MC1R also plays a role in adrenaline production, and it has been suggested that redheads not only produce more adrenaline but that their systems can access it more speedily, too—in other words, they fire up more rapidly than others (that fear/flight response). Certainly Hans von Hentig, writing in the *Journal of Criminal Law and Criminology* in 1947, thought so. Tracing the history of the outlaw in the West from 1800, he lists examples ranging from "Big Harpe," who terrorized the Ohio Valley in the early nineteenth century and who had, apparently, "coarse hair of a fiery redness" to Wild Bill Hickok ("long auburn hair hanging down over his massive shoulders"). Von Hentig adds to the list of redheaded no-goods Sam Browne and Jesse James, speculating that the appearance of these men was remembered because with the color of their hair, they stood out, which is the sort of logic that sounds a little as if it might be getting into the all-too-familiar area of "they acted like redheads, so must have had red hair"; but he does also observe that "redheadedness also is often combined with acceleratedness of motor innervation," which, medically speaking, would seem to imply that redheads exist in a permanent state of heightened stimulation. I'm not so sure about that—but these were all gunmen, quick on the draw.

And suggestively, in light of this, science has very recently begun to explore a link between red hair and Tourette's syndrome.

Tourette's syndrome (Ts) is an astonishingly complicated, interwoven set of symptoms that can range from small tics and mannerisms to involuntary outbursts of obscene vocabulary. It's thought that Mozart may have been a sufferer. Technically it is termed a chronic, idiopathic, neuropsychiatric disorder, which should warn you that at present, almost nothing can be categorically ruled out as a cause. In 2009 an Australian pediatrician, Katy Sterling-Levis, whose thirteen-year-old son had just been diagnosed with the disorder, was attending a conference on the syndrome and was struck by the number of redheads in the room—or how they were "overrepresented," to use the unemotive language of the scientific study. The pattern of inheritance for Ts is autosomnal recessive. In other words, two copies of the gene must be present for the condition to result, just the same as with red hair. A survey organized to test the strength of the connection revealed that 13 percent of those with Ts had red hair, compared to the normal average within the population of 2 to 6 percent.[101] More than half of the sufferers of Ts had one or more relatives with red hair. ADHD and hyperactivity are also associated with Ts, and a few pediatricians have drawn a much-contested connection between the pale skin/red hair phenotype and hyperactivity, to boot. In adults as in children this is characterized (among other symptoms) by impulsive and inappropriate behavior, mood swings, racing thoughts, and a craving for excitement. Spencer Tracy, one of Hollywood's leading men in its Golden Age, and one of the very few actors to make it seriously big despite an unmistakable tawniness in his natural coloring, was hyperactive as a child. It's also striking how many symptoms of hyperactivity seem to have been present in the poet Algernon Swinburne's behavior.

Swinburne was born in 1837. His mother was a daughter of the third Earl of Ashburnham, and Swinburne proudly traced his red hair back to Henry VIII. His well-to-do family sent him to Eton, where his cousin Lord Redesdale left this toe-curling description of him:

> What a fragile little creature he seemed as he stood there between his father and mother, with his wondering eyes fixed upon me!. . . His limbs were small and delicate; and his sloping shoulders looked far too weak to carry his great head, the size of which was exaggerated by the tousled mass of red hair standing almost at right angles to it. Hero-worshippers talk of his hair as having been a "golden aureole." At that time there was nothing golden about it. Red, violent, aggressive red it was, unmistakeable, unpoetical carrots. . . . His features were small and beautiful. . . . His skin was very white—not unhealthy but a transparent tinted white, such as one sees in the

---

101 Katy Sterling-Levis and Katrina Williams, "What Is the Connection between Red Hair and Tourette Syndrome?," *Medical Hypotheses*, 73, no. 5 (November 2009): 849–853.

petals of some roses. . . . Altogether my recollection of him in those school days is that of a fascinating, most loveable little fellow.[102]

The fascinating, loveable little fellow went on to scandalize the London literary scene with his poetry (technically some of the most accomplished in existence, still, but in its subject matter suggestive, erotic, and deviant, at the height, in the 1860s, of Victorian respectability), and outrage London society with his behavior as soon as he could.[103]

Swinburne's endlessly fluttering hands and feet had been diagnosed in childhood as "an excess of electrical vitality." As an adult he seems to have been willing to do anything to make himself the center of attention—that stunt of sliding nude down the bannisters at Rossetti's was one of many. He rapidly became both an alcoholic and a user of opium, and he posed (according to Oscar Wilde, whose gaydar, one would think, would be faultless) as a homosexual.[104] Medical opinion now is that he may well have suffered some degree of brain damage at birth and was possibly hydrocephalic. Such was his notoriety that when on July 10, 1868, he fainted in the Reading Room of the British Museum, the event made the papers. He filled perhaps something of the same role as the comedian Carrot Top does today—determined to be weird, and if possible, to make a good

102 Quoted in Christopher Hollis, *Eton: A History* (London: Hollis and Carter, 1960), 291–2.

103 Leonard Shengold, *op. cit.*, in his chapter on Swinburne makes a sympathetic and insightful case in his defense.

104 Wilde said of Swinburne that he was "a braggart in matters of vice, who had done everything he could to convince his fellow citizens of his homosexuality and bestiality without being in the slightest degree a homosexual or a bestialiser."

living out of it—but you do wonder with both what the image of the redheaded man on either side of the Atlantic might have been without them.

It seems almost a shame. Genetic science will one day solve so many of the mysteries that orbit the edge of our imaginations. The truth, or otherwise, of the "blond Eskimos" of Coronation Island in northern Canada, reported as recently as the early twentieth century, will be pinned down, one way or the other; the reconstructed faces of the Tarim mummies, so different from their Han Chinese neighbors, will be allied to their genome and end once and for all speculation as to whether they were European, Asian, or a coming-together of both. The capital of the Budini will be located; and the Udmurt people will be incontrovertibly assigned the same region genetically as they now inhabit geographically, poetically described as being between the Great Forest and the Great Steppe. Their language will give up its last Finno-Ugric secrets, and John Munro will rest easier for knowing he got one thing right: "A white freckled skin, greenish eyes and fiery red hair are characteristic of the Finns, Rühs, and other people of the Baltic highlands." The enormous graveyard of maybe a million mummies sitting on the edge of Fag el-Gamous, south of Cairo, will provide an answer as to why they seem to have been buried with blonds in one area and redheads in another, and who these blonds and redheads in first to seventh century AD Egypt were in the first place.[105] And (which would be particularly handy in this case) a simple means will be discovered of

---

105 See http://www.livescience.com/49147-egyptian-cemetery-million-mummies.html.

distinguishing true red hair in life from hair whose color in death has been altered by acids in the soil or fungi or bacterial decay. The reason for the red *scoria*-stone topknots on the statues, or *moai*, of Easter Island in the southeast Pacific will be found, and their links (or lack therof) to the cult of the redheaded Birdman that persisted right into the 1860s will be explained. An unguessable amount of the history and mythology of Easter Island has been lost to us, but it's hard not to see the use of red as connected in some way to those red-topped statues, gazing out over the Pacific for so many centuries before. The startling possibility that the first Maori settlers in New Zealand may have found a native people already living there, the Ngāti Hotu, who had fair skin, green eyes, and red hair, will be confirmed one way or the other.[106] And we may then be closer to understanding why it is that Edward Tregear (1846–1931), one of the first to study the Maori, found so many terms among Polynesian speakers to describe red hair:

> Samoan, *'efu*, reddish-brown: Tahitian, *ehu*, red or sandy-colored, of the hair, *roureuhu*, reddish or sandy hair: Hawaiian, *ehu*, red or sandy hair, ruddy, florid; *ehuahiahi*, the red of the evening, or old age; *ehukakahiaka*, the red of the morning, or youth . . . Marquesan *hokehu*, red hair.[107]

And finally, the coppery topknot of the elongated Paracas skull and the skull itself, and its fellows, will be tested by a lab anybody has ever heard of, and be returned from the star-gazing world of

---

106 Kerry R. Bolton, "Enigma of the Ngati Hotu," *Antrocom Online Journal of Anthropology*, 6, no. 2 (2010): 221–26.

107 Quoted in Kerry R. Bolton, *op. cit.*

Indiana Jones and the Kingdom of the Crystal Skull to this small blue planet here.

When I was very young, but not too young to have noticed that I was the only redhead at my village school, I aligned myself, as many a redhead has done before, with a Celtic ancestry. I was convinced I must have much Irish and some Scottish blood to boot, and that was where the red hair came from. My mother had a vague idea that there was Irish blood in her family, and my father's family names were Ewart and Colliss, one of which was definitely Scottish, and the other of which I decided was also going to be Scottish, to keep things tidy. Besides, all redheads have Scottish or Irish blood, right? Show me a writer who won't take the nice smooth plotline over the untidy truth, and who doesn't start off by rewriting their own identity in some way. Anyway, I liked the idea of my Celticness; I thought it made me different from my friends and superior in some way to my enemies; and no one contradicted me, because obviously I had the red hair to back it up. But in one of those everyday marvels of our age, of course you can now test these assumptions out, with a DNA kit. So I now know that I appear to be 100 percent *un*-Celtic, that my haplogroup, centered in the North of England, is 46 percent Brit; 38 percent European (centered—who knew?—on Switzerland); 13 percent Scandinavian—Norwegian, by the looks of it; and 3 percent Eastern Middle East. I might as well be that Ur-redhead.

Just about everyone on the planet will have some DNA from the Middle East. We came if not from there then through there, and the haplotype will be there still. Norway and the North of England connect, via the Vikings. Switzerland, I have no idea. But according to the company who did the analysis, there is, on the database of this one company, twenty-seven pages' worth of people who have also had their DNA tested, and to whom I am genetically related in some way. This is not as revelatory as it may sound—you only have to go back six centuries or so and statistically there would be a European someone to whom everyone now alive in Europe would be related. And of course the analysis doesn't tell me how many of those twenty-seven pages' worth of folk have red hair. But there is a pleasing paradox here—that a science that begins by measuring difference ends by making brothers and sisters of us all.

So that is what I am in terms of cells and ancestors piled one atop the other. But to sort out what all that means in terms of being a redhead today, you need not the hard-and-fast of science but the loose-and-sloppy of everything else.

FREAKS OF FASHION

There is a deep-rooted and unaccountable
prejudice against this much-abused shade of colour,
which it is quite possible some unexpected
freak of fashion may one day change.

T. F. THISELTON-DYER,
FOLKLORE OF WOMEN, 1906

Every autumn it's the same—the days grow short, the skies are gray, the evenings dark, the nights too long. Leaves pile against the railings and crunch underfoot, and then are gone, and with them, color abandons the Northern hemisphere. The eye craves it, just as did the eyes of our ancestors before us, decorating their tombs with red-haired goddesses to give us life again, to bring back the sun. The glow of bonfires and the sparkle of fireworks across the UK in November doesn't only celebrate the last-minute discovery of Guy Fawkes ("a tall, powerfully built man, with thick reddish-brown hair, flowing moustache, and a bushy reddish-brown beard," according to the historian Antonia Fraser) and his barrels of gunpowder, and the foiling of the Gunpowder Plot, they also echo the festivals of All Hallows and of Halloween, of Diwali, the Day of the Dead, Hogmanay, and the world-turned-upside-down of the Roman

Saturnalia. We all feel it, that ancient winter hunger for warmth, for light; and every autumn, in answer, the red carpet—where else?—fills up with celebrities sporting newly dyed red hair. As *Vogue* puts it, "Mythologized, demonized, celebrated: every shade, from carrot to scarlet, conveys an inscrutable allure."[108] A wholly unscientifically assembled list includes, as I write, Jena Malone, Kirsten Dunst, Amy Childs, the models Suki Waterhouse and Karlie Kloss, Amy Adams (a great spokeswoman for the advantages of going red), Sofia Vergara, onetime Bond girl Olga Kurylenko ("It's been red for a couple of months . . . I'm definitely getting more attention from men"), Katy Perry, and Katie Yeager. Katie Holmes announces that she's always yearned for red hair, and the world stumbles to a halt. Meantime, in a pleasing twist on the notion of the undead redhead, a warmth of tone has been noted in the hair of the actor Robert Pattinson, Edward in the *Twilight* series. Nothing captures the eye like red hair.

We know we are drawn to the color red. Waitresses dressed in red supposedly make more in tips. Men are meant to find women more attractive if they wear red. Women are meant to find images of men more attractive when they are shown those images against a red background. And there is of course the age-old sexual allure of the flame-haired temptress, something neither Hollywood nor the rest of the celebrity industry needs to be told. But the media company Upstream Analysis suggests another reason for our fascination with red hair. As well as drawing the eye and grabbing an audience's

---

108 *Vogue*, "The Best Redheads," July 2014. See http://www.vogue.com/946492/best-redheads-jessica-chastain-amy-adams-julianne-moore-and-more.

attention, its rarity, Upstream suggests, sparks the reward-seeking instinct in us, firing up the center of our brain that is most highly sensitized to novelty.[109] When we see something unusual, we want to get close to it. We want to inspect it, or engage with it in some way, or even to touch it. In other words, we buy into red hair. This is the reason, Upstream concludes, why up to a third of all TV advertising features a redheaded character, when the actual percentage in the population hovers at a scarce (and thus noteworthy) 2 to 6 percent. If nothing draws the eye like a redhead, it would seem that nothing sells like one, either.

Every mother of every redhead the world over will know the experience of complete strangers coming up to comment upon or even to reach out and touch their children's hair. Growing up as a redhead, it sometimes felt as if the last person my red hair belonged to was me—the person from whose scalp it sprang. It was one of the many things that made growing up as a redhead so deeply confusing. Grant McCracken again: "Redheads become a handy plinth, a medium for the message, a carrier of the color." Once again, the hair overpowers everything else. It becomes all people see. The normal barrier, the invisible area around us that we all own, and into which others do not enter without our permission, apparently doesn't exist if your hair is red, and you're too small for your wishes to count. (And this we endure as children on top of our heightened flight response and hair-trigger adrenaline production. So next time you are tempted, all I'm saying is, *ask*.) People—like Rachel Lynde, in

---

109 See http://www.theatlantic.com/business/archive/2014/08/redheads-are-more-common-in-commercials-than-in-real-life/375868.

her encounter with Anne Shirley—talk about and comment upon your hair while you yourself are standing there beneath it, as if you were merely wearing it, like some kind of hat. And if, tiring of this as you grow older, as many a redhead does, and as your dominion over your body increases, you should cut your red hair, or dye it, there is outrage, as if the thing you had changed was everyone else's property, which you have damaged, willfully. These are such common behaviors in the non-redheaded world, they're convincing evidence that Upstream Analysis is on to something here.

And such primates are we that anything we admire we wish to emulate. The human race has been changing the color of its hair with henna, walnut juice, saffron, red wine, red ochre, lye, vitriol, indigo, and woad for millennia, but in the twentieth century the first commercially available hair dyes created an entire new industry and revolutionized society's response to dyed hair into the bargain. You can't emulate an image without there being one to emulate, so hand-in-hand, or rather chicken-and-egg with this, went Hollywood's remaking and shiny repackaging of the actresses who entered its studio system, none of which would or could have happened as it did without one Maksymilian Faktorowicz, or Max Factor, as he is better known.

Max Factor's life really should be made into a movie itself. One-time hairdresser to the Russian court—so prized that he conducted almost all his early professional life under guard—he fled Russia to hide the fact of a marriage for which he had failed to receive the tsar's permission, using his own makeup to fake the symptoms of jaundice in order to be allowed to leave Moscow. Arriving in America in 1904 with wife and very young family in tow, he was promptly fleeced of most of his savings by his partner in the Louisiana World's Fair, and then found his doorstep darkened by his ne'er-do-well half brother, John Jacob, or "Jake the Barber," to give him his mob name, a Prohibition gangster and con man who did once, literally, break the bank at Monte Carlo. Unsurprisingly, Max kept moving, heading west and arriving in Los Angeles in 1908 and setting up a business hiring out wigs to bedeck Hollywood's film extras. Then he did the extras' makeup, too. Then he did the makeup for the stars, as well. Then he invented the term "makeup." Within an impressively short span of time, the name Max Factor and the glamour of Hollywood had become just about synonymous. In 1935 Max opened the "Max Factor Make-Up Studio" (note the word "studio" in the title), with its color-coded makeup salons: peach for "brownettes" (a type that failed to catch on, and a rare example of a Max Factor marketing misstep); pink for brunettes, powder-blue for blondes, and a pale mint green for redheads (Fig. 30).[110] This Redheads Only room was officially opened by Ginger Rogers, and in painting it mint green Max created a color association that bedevils redheads to this day, and one that Maria Oakey would have taken issue with for sure: "Wherever there is red in the composition of the hair, green (*not* a pale green, which should only be worn by blondes). . . will be becoming." But no matter, Max's successes far outnumber the fails. He created Clara Bow's heart-shaped lips, and Jean Harlow's platinum blondness (if you thought those Renaissance hair dyes sounded undesirable, Jean Harlow's candle-flame white was achieved with a weekly wash of

110 Fred E. Basten, *Max Factor: The Man Who Changed the Faces of the World* (New York: Arcade Publishing, 2012).

ammonia, Clorox, and Lux soap flakes). And although the timing of the claim is hard to be exact about, by repute, Max also created Rita Hayworth's tumbling copper curls.

Hollywood has been graced by many a redheaded beauty, but nearly thirty years after her death, more than forty since her tragically early retirement, the first name that comes to anyone's lips when you put the words "Hollywood" and "redhead" together is Rita Hayworth. The second is *Gilda*, the part she plays in the film of the same name. The glorious, sensuous Gilda, undulating across the stage as she sings "Put the Blame on Mame" in elbow-length black satin gloves and a dress that required the real Rita Hayworth to wear both a corset to get into it and a hidden harness to keep it in place. The third name is of course Lucille Ball. Neither Rita nor Lucille was genetically a redhead, any more than Debra Messing or Christina Hendricks were born with the red hair they have worn so proudly, but no matter. All four, and many more, are a part of the image of the redhead as we all receive and respond to it today; what makes Rita Hayworth and Lucille Ball so extraordinary is that they managed to be so in monochrome. *Gilda* was filmed in 1945, when the film industry was still veering between the novelties of Technicolor and good old black and white. Lucille Ball's TV show (with its endnote sign-off "Make-up by Max Factor") went on the air in 1951. Color TV sets—or shows—did not exist in any numbers until the mid-1960s. What miles we have covered since then.

Here's a case history to ponder. In 1941, Rita Hayworth, coming up to the peak of her career, was in two movies, one—*Blood and Sand*, a cautionary tale of how badly bullfighting and hubris go together—in Technicolor; the other, *The Strawberry Blonde*, paradoxically, in black and white. In *Blood and Sand*, in color, and playing on her own "exotic" Mexican ancestry, Rita plays a socialite, Doña Sol, a sexual aggressor, the possessor of a kind of fantasy sexuality, as it is described by her biographer Barbara Leaming.[111] Darryl Zanuck, the film's producer, had originally wanted Carole Landis for the role, but Landis refused to dye her hair red for it, hence the part went to Rita. This gives some idea of the importance given to the aesthetic of Technicolor in the movie (wrong hair and you're out), with sets that were meant to evoke the works of the Spanish painters El Greco, Goya, and Velázquez. But you don't get the impression much thought went into the character of Doña Sol, who seems to be driven by nothing—albeit driven very beautifully by nothing—other than an urge to add notches to her bedpost. In *Strawberry Blonde*, Rita is a society girl again, but the all-American version thereof, and there's a neat if slightly queasy-making piece of story-telling/stereotyping in this film. As the cast list comes up on the screen, James Cagney is introduced as "Biff Grimes," Olivia de Havilland as "the girl he adores." When Rita comes up on the screen the line under her name is simply "M-m-m-m-m-m." She's not a character with a story, she's simply the audience's lip-smacking reaction; not a human being but a symbol, a vector for desire.

And ultimately, I would argue, one that is a dead end. The image of Gilda has been referenced in film after film, but in the end, what can you do with a sex symbol other than make it sexier and sexier, until it becomes a parody of itself? Once Jessica Rabbit, who

---

111 Barbara Leaming, *If This Was Happiness* (London: Sphere Books, 1990).

famously isn't bad—"just drawn that way"—had swung onto the screen in the film *Who Framed Roger Rabbit?* (1988), we'd gone as far as we could go. The association between redheaded women and sex has become so knee-jerk that advertising now doesn't even need a body for its message to be understood. In 2012 Sleepy's, the US mattress company, launched a range of gel-infused mattresses, guaranteed to keep you cool in the sweltering New York summer. The advertising image chosen was a sparkling white duvet, a pillow, and between the two a woman's head of rich red curls. All you saw of her was her hair and an ear. Sleepy's tag line was simply "HOT IN BED?" You start to understand Rita Hayworth's sad complaint that "men go to bed with Gilda, but they wake up with me." We are not all hot in bed. We do not all, always, want to be. Centuries and centuries of stereotyping, of social and sexual expectation, decades of advertising, can really make you feel the cold.

And did those same ads for *Strawberry Blonde* feature the redheadedness of Rita Hayworth's costar, James Cagney? No, they did not. In all this, where were the redheaded men?

Cagney had been a significant star since the 1930s, when Lincoln Kirstein described him as "a short redheaded Irishman, quick to wrath, humorous, articulate in anger . . . the semi-literate lower middle-class . . . mick Irish," and the *L.A. Record* as a "red-haired

Bowery Boy . . . fiery-tempered, but with a warm Celtic smile."[112] Note the "but," implied in one description, explicit in the other. We are being invited to like and admire Cagney despite the flaws in his presented image—his red hair being but one of them. We are to overlook that—in fact in the posters for *Strawberry Blonde*, his hair is brown. You can make an argument, however, that Cagney's career and the parts he played track the transformation of the pugnacious Paddy, the brawler, the laborer, the Celtic cliché of the nineteenth century into the twentieth century's sleek urban animal, albeit one who in Cagney's movies is very often on the wrong side of the law, ready and able to turn on that irresistible Celtic charm whenever needed. By the time Errol Flynn came along (chestnut-haired in life, since you ask) playing the wild Irish card could be the basis, almost, of an entire career. Again, we are to overlook that tint, or taint, of red. Another wholly unscientific list: Eric Stoltz, Ewan McGregor, David Caruso, Rupert Grint, Damian Lewis, and Benedict Cumberbatch. Simon Pegg says he is not; Michael Fassbender happily declares himself a ginger Viking, and who are we to argue? But unlike Rita, Lucille, Debra, or Christina, not one of these has dyed his hair red as a means to further his career; indeed Benedict Cumberbatch, deeply annoyingly, in his best-known role as Sherlock Holmes, has his dyed almost black—deeply annoying in that if ever there was a character who could and should be played as a redhead, it's the eccentric, intellectual, unconventional, alienated Sherlock. When the photographer Thomas Knights launched his *RED HOT*

---

112 Quoted in Ruth Barton, *op. cit.*

100 project in 2014, photographing redheaded men—staggeringly handsome redheaded men, fair enough—and encouraged them to talk about their experiences of going through life with red hair, one of his sitters confessed that he had found it easier to come out as gay than to come out as a redhead. What does that say about red hair as one of the last great social prejudices? It is, still, different for girls, but here is another paradox: we don't judge or value men by their appearance in the same way as we judge or value women. So whereas redheaded actors, with some clever manipulation by their image-makers, may escape the stereotyping that comes with red hair, and play against it, and have us overlook the color of their hair entirely in our approval of them, redheaded women, whether in the public eye or no, are still bound by stereotypes stiff as a straitjacket. As an adult it can feel as if you simply get to pick which one you'll don, out of a choice of three: Rita, Scully, or Lucille. Without Lucille Ball, that choice might have come down to just two.

Lucille Ball single-handedly created the third modern variant in the palette of redhead female types. Her character in *I Love Lucy* is a kook, a kind of back-formation from the winning, redheaded little girl characters of children's fiction at the start of the twentieth century, only this time in the body of a grown woman; a housewife who dreams of stardom, who talks fondly of her "henna rinse" on the air, who is unreliable and unpredictable, but human and believable and irresistible because of it. In real life Lucille Ball had her signature hair color created by her hairstylist, who declared "the hair may be brown but the soul is on fire!" Off set she was one of the canniest and most undauntable female studio executives there has ever been. She paid a price, of course, in chauvinism and name-calling, but if

Fig. 25: Edgar Degas, *La Coiffure, c.* 1896. Is this a servant preparing her prostitute mistress for the night's work, or an aristocratic young woman being tended to by her maid? Are the two mother and daughter? The one is impassive, absorbed in her task; the other holds on to the roots of her hair with one hand and raises her other in pain. Is the interior where this scene takes place a boudoir, or a brothel? Only one thing is certain: it's all about the hair.

Fig. 26: Cora Pearl (born Emma Elizabeth Crouch) as Cupid in Offenbach's *Orpheus in the Underworld*, of 1867. The operetta was notorious for its cancan; Cora by this time had also appeared in public as Eve, and at a dinner party at her château in the Loiret, had herself served naked on a silver platter.

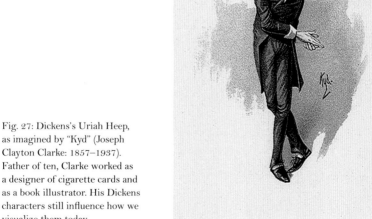

Fig. 27: Dickens's Uriah Heep, as imagined by "Kyd" (Joseph Clayton Clarke: 1857–1937). Father of ten, Clarke worked as a designer of cigarette cards and as a book illustrator. His Dickens characters still influence how we visualize them today.

Fig. 28: Tintin, who first appeared in
1929. His creator, Hergé, may have been
influenced by a freckle-faced red-haired
Danish fifteen-year-old, Palle Huld,
who in 1928 won a competition to travel
around the world, Phineas Fogg–style,
and who completed the trip in just
forty-four days.

Fig. 29: Around 26 percent of the population of the Solomon Islands carry a unique gene found nowhere else on earth. This gives 5 to 10 percent of them fair hair, which ranges from bright blond to a light ginger.

Fig. 30: A detail of the decoration of the Max Factor mint-green "Redheads Only" room at the Hollywood History Museum, with a Max Factor advertisement endorsed by none other than Lucille Ball.

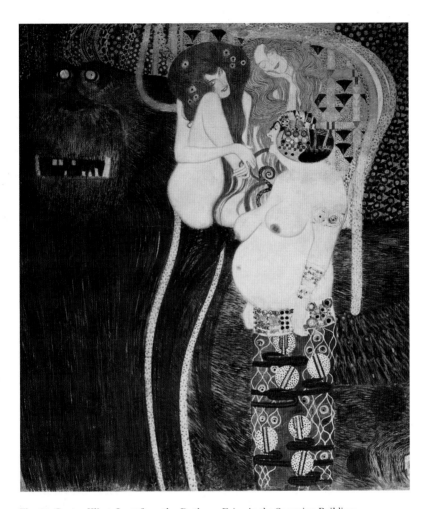

Fig. 31: Gustav Klimt, Lust, from the *Beethoven Frieze* in the Secession Building, an exhibition hall built in 1897 in Vienna. The building was created to show the work of artists who had seceded from the Austrian establishment, and was financed by Karl Wittgenstein, a steel tycoon and father of the philosopher, Ludwig Wittgenstein. The plump dark-haired figure is Intemperance; the sleeping blonde represents Debauchery.

Fig. 32: Redhead apocalypse. A detail of the group shot from Redhead Days in Breda in 2014. No fewer than six thousand redheads from all over the planet attended the festival.

Fig. 33: Sterra Vlamings, a redheaded model of Dutch and Senegalese descent.

the soul was on fire, the business brain beneath the red hair was second to none, and she knew it. It's more than a little ironic therefore that the character she created—zany, unintellectual, and delightfully flawed—should be what she has been remembered for, referenced by innumerable actresses ever since: Debra Messing's Grace in *Will & Grace*, for one; Alyson Hannigan's character Lily Aldrin in *How I Met Your Mother*, for another—women whose comedy flows from the gap between what they try to be and what they are. And because she was flawed, and warm, and human, Lucy, the original, was within the reach of housewives, mothers, women everywhere. Hair dye would not work the miracle that might turn the average woman into Rita Hayworth. It might not give you the go-getting career of redheaded *Brenda Starr, Reporter*, who was at the height of her comic-strip fame in the 1950s. But it could bring you closer to the more achievable glamour of Lucille Ball. And we *knew* Lucy dyed her hair, and wore makeup, because in her show she told us so. No more the scandalized reaction of Aunt Marilla to Anne Shirley dyeing her hair ("A wicked thing to do!"). If it was okay for Lucy, it was acceptable for every woman. Her signature look even crossed the Atlantic into my parents' living room, with its shoe box–size television screen. And where demand leads, the advertising men are sure to follow. Except of course that by this time, some of those men were women. And one of those women was Shirley Polykoff.

In 1955 Polykoff was working at the ad agency Foote, Cone & Belding and took over the Clairol account, for whom she created the "Does she or doesn't she?" tag line for their range of do-it-at-home hair dyes.[113] What's particularly pleasing about this bit of advertising genius is the fact that Polykoff, who was born in Brooklyn in 1908, was Jewish and based her slogan on a perfect bit of Yiddisher-momma speak: *"Is* she or *isn't* she?" (complete with hand gestures). Thus one strong woman channeled generations of others, and changed women's relationship to their hair forever. If you wished to be a blonde, a brunette, or a redhead, now you could be, with a use-at-home hair dye that gave results so natural-looking, said the slogan, that your secret would be safe. And why would you do this? Because you're worth it—the second game-changing slogan in the field.

The original version of this, created by another woman advertising executive, Ilon Specht, for L'Oréal in 1973, was "Because *I'm* worth it"—an important difference to the warmer, fuzzier, we're-all-in-this-together variant used by L'Oréal today. "Because we're worth it" is several steps on from the hotly contested feminist debates of the 1970s. Of course we're worth it. We *know* we are.

And once that ball was rolling, once the products had been created that made dyeing your hair at home your choice, and so easy, and so commonplace, and so much something the stars in Hollywood were doing too, all the age-old censure was gone. An industry that had been worth $25 million became worth $200 million a year. The two shades of red offered by L'Oréal in 1970 were sixteen by 1989, with Redken offering twenty-nine and Clairol forty-three. In fact, it has been estimated that more red hair

113 Malcolm Gladwell, "True Colors," *The New Yorker,* March 22, 1999. See http://gladwell.com/true-colors.

dye is sold per annum off the shelves of supermarkets and pharmacies worldwide than any other shades.

Mrs. Roger Rabbit shares her ancestry between Rita Hayworth and another 'toon, the "little redheaded ball of fire, Red Hot Riding Hood," of 1943, with her floop of red hair over one eye, tiny scarlet skater's dress, and a Katharine Hepburn cut-glass accent. *Red Hot Riding Hood* is still considered one of the greatest cartoons of all time and has the rare distinction of having brought down the wrath of the censor on the director, Tex Avery, when the wolf, a night-club lothario (Grandma in the same cartoon becomes a frisky Park Avenue matron) became just a little *too* excited by Red Hot Riding Hood's performance. She was also supposedly influenced by the real-life Hollywood legend Lana Turner, a dazzling blonde. In the movies, if you want to show a blonde gone bad, you turn her into a redhead. It was done to Jean Harlow, of all women (of all blondes), in *Red-Headed Woman* in 1932. Harlow's character in the movie, wearing a flame-red wig throughout, was a home-wrecker, blackmailer, adulteress, and would-be murderess, thus ticking just about all the boxes for the sinful female redhead. The same transformation is still being used today in *The X-Men*—when Raven is Raven, she's a blonde. When she's the villainess Mystique, her hair turns copper.

This is a version of the redhead—the evil redhead—that began with Lilith. This is red hair as a marker for sorcery and the supernatural, nowadays with a pinch of existential angst thrown into the mix as well. Possibly one of the best and at the same time one of the most delectably evil redheads was created back in 1866 by Wilkie Collins, in the character of Lydia Gwilt in his Victorian mystery *Armadale*. Lydia is not only a proto-vamp, she seems to steal something of the vampire's eternal youth as well, seducing men much younger than her own advanced age of (God help us) thirty-five. Wilkie at this date was forty-two. The book kicks off with the deathbed confession of a murderer, and by the standards of the 1860s gets only more sensational from there on in. Here is Collins introducing his antiheroine:

> This woman's hair, superbly luxuriant in its growth, was the one unpardonable, remarkable shade of colour which the prejudice of the Northern nations never entirely forgives—it was red.

Never entirely forgives, eh? The description goes on:

> Her eyebrows at once strongly and delicately marked were a shade darker than her hair [every redhead knows the redhead eyebrow dilemma, of brows so fair they disappear; Lydia however is so flawless that she is spared it], her eyes large, bright and well-opened were of that purely blue colour. . . . Her complexion was the lovely complexion which accompanies such hair as hers—so delicately bright in its rosier tints, so warmly and softly white in its gentler gradations of colour on the forehead and neck.

In the complete description, we get her mouth and nose as well as her forehead, eyes, and neck, against which one can only imagine Collins resting, in his imagination, once his delineation of Lydia is complete, with a happy sigh. Lydia is poison, but she is the kind her victims imbibe willingly, and Collins's pen relishes every detail of her. She is a fortune-hunter, a seductress, possibly a bigamist, a murderess, and, ultimately, a suicide. This is the evil redhead, and they are never less than gorgeous, and always deadly in a particularly sexy way. Think of Edvard Munch's *Vampire*, of 1893, a painting the artist entitled *Love and Pain*, but his public, viewing the painting, took one look at the female figure nuzzling down into the back of her lover's neck, with her red hair dripping down around them both, and retitled it for him. Or there is Gustav Klimt's figure of redheaded Lust, in his *Beethoven Frieze* in Vienna of 1902, her long red hair curling between her thighs in a preposterously overt sexualizing of the much more modest pose, five hundred years before, of Botticelli's redheaded Venus, her head tilted, her predatory gaze measuring up the viewer (Fig. 31). There is Bevis Winters's 1948 hard-boiled gumshoe Al Rankin, encountering the redheaded Maisie Tewnham (also a bigamist) in *Redheads Are Poison*: "Her bright red hair ought to warn those foolish guys not to succumb . . . " The real Bevis Winter was an Australian, living in the genteel English seaside resort of Hove. In the same scarlet vein, there is Poison Ivy with her toxic kiss, who squares up even to Batman. Or even Bree Van de Kamp in *Desperate Housewives*, the perfect housewife and mother, with her spotless home and flawless hair and makeup—and her equally perfect willingness to kill to keep it that way. Her red hair is the viewer's clue that all with this woman is not as it seems.

Of course transformation works both ways. If you want to tame a redhead, turn her into a blonde. Orson Welles did this to Rita Hayworth when in 1947 he cast her as Elsa, the eponymous lady in *The Lady from Shanghai*. The film is heavy-handedly noir, and like many an Orson Welles movie, one of the most intriguing things about it is the way it pulls itself apart. Why Welles should have thought red hair unsuitable for a woman who manipulates every man around her is a mystery. Possibly in typical Wellesian fashion he was interested in defying one stereotype to remake another. Or, less pleasantly, it's a symptom of an attempt to control Hayworth as his marriage to her failed. But other actresses have done the same: Gillian Anderson has reverted to her natural blonde since playing Agent Scully—an excellent way to break the hold of that character on her career, perhaps. Christina Hendricks did the same to mark the end of *Mad Men*. Nicole Kidman, to the distress of many another female redhead out there, claims blondeness for herself these days, too. The red-to-blond and blond-to-red toggle switch: it comes in very handy. It was used by Disney in 1989, when their *Little Mermaid*, Ariel, was given red hair to differentiate her from Daryl Hannah's blonde Madison, the mermaid in 1984's *Splash*. A variant was employed in the 1960s sitcom *Gilligan's Island* in the opposing female types of Ginger and Mary Ann, one a redhead, one a brunette.

There is of course an argument, basically a chauvinistic argument, that changing your hair or hair color or lipstick or winter coat is proof that women are at the hopeless mercy of whatever the advertising industry last told them they must wear or be or do. There is in my view a much stronger argument that taking control of your appearance (your *body*), and having the freedom to make choices about it should be a part of the life of every human being on the planet, particularly every woman, and is a part of the ongoing emancipation of the female sex and, now, many a onetime minority group as well—redheads included.

Some years ago I spent a fortnight in the Ukraine, around the town of Sebastopol, famous as a locus of the Crimean War of 1853–6.

Much of Sebastopol looked as if the shelling from that conflict had only just ceased. This had clearly been a very tough place to live for a very long time. Almost every evening there would be a power cut, just to remind the Ukrainian people that Mother Russia had her finger on the switch. President Yushchenko was making his first public reappearances, still as heavyweight and charismatic a figure as ever, albeit pocked and scarred from someone's attempt to poison him with dioxin. The most popular fashion among the men was for black plastic peacoats, reaching to their knuckles and their knees, which gave them all the appearance of being low-ranking Myrmidons among the ranks of the KGB (for all I knew, from within my gaggle of Western tourists, they were). Far and away the most popular fashion among the young women of Sebastopol was for zinging, cranberry-colored hair, or orange of a beacon-like brightness, worn with red-carpet makeup applied with a professionalism that would have had Max Factor weeping tears of joy, and with vertiginous, stiletto-heeled boots and pencil skirts that had their wearers sashaying like Marilyn Monroe (who began life as a strawberry blonde, if her famous "Red Velvet" calendar shots of 1949 are to be believed). In this performance of hyper-femininity, these shining orbs of red or orange would go bobbing down the streets as their owners picked their way around potholes in the pavements and missing cobblestones in the roads. Talking to them, it was clear that this was the way they chose to counter the uncertainties and hardships of their lives, by expressing the same kind of intense, heightened femininity previously given shape in the "New Look" by Christian Dior, in the drab, gray, ration-regulated world of Europe post-1945. And if you are going to be as powerfully female, as ultra-feminine, as you possibly can, it seems the hair color you choose is red. Many Ukrainians have Tartar ancestry and dark coloring, and it is only human nature to want the opposite of what you've got, so you might imagine the hair color they would most desire would be blond, but not in this case. Not for the redheads of Sebastopol. For them, red was the only color that declared their pride in their gender, their defiance of all life might throw at them, and their solidarity with one another. Both the orange and the red were completely elective (there was no possibility of mistaking either color for anything found in nature) and thus said these woman not only had this choice but the freedom to exercise it, too. For the redheads of Sebastopol, red was the color of empowerment.

One of Shirley Polykoff's ads for Clairol showed a redheaded mother and daughter, the idea being that the mother's hair color would look as natural as the daughter's, while the red of their hair established for their audience the familial bond. In the movie *Atonement* (2007), for the wedding scene, the director filled one entire side of the church with redheads (that's me in the pale blue suit, dead animal over one shoulder, handsome RAF husband at my side), as a visual shorthand to establish that this was the family of the twin redheaded boys so central to the plot. Two redheads together find that people always assume they must be related to each other in some way. Evian mineral water made use of this idea in one of the recent ads in its "Live Young" campaign, with a pale-skinned, anxious-looking, red-haired young man twinned with pale-skinned, anxious-looking baby with a cockatoo quiff of orange curls. Elle Fanning and Naomi Watts, playing mother and daughter in the film *Three Generations*, are another example, with the truly red Susan Sarandon joining them in the role of grandmother. Even earlier: there is a story that Rossetti once took Lizzie Siddal and Algernon Swinburne to the theatre together. The boy selling programs was already unnerved by Lizzie's pallor and above all by her hair. Reaching the end of the row and encountering Swinburne, he is supposed to have dropped his armful of programs to the floor and let out a screech: "Here's another of 'em!" We crop up so unexpectedly that the non-redheaded world sometimes seems to find it easiest to assume we all belong together, that we must all be family, in some way. Which throws an interesting light on a recent change in attitudes toward red hair.

In 2011 the world's largest sperm bank, Cryos International (which, ironically enough, is based in the redhead-rich land of Denmark) announced that it would no longer take deposits (their word, not mine) from redheaded donors. This in turn inspired the Italian photographer Marina Rosso to create a photographic matrix of forty-eight different types of redhead, published as *The Beautiful Gene*. Rosso's starkly lit portraits, arranged according to the criteria most commonly used by clients of the sperm bank, stare back from page after page—the human race in all its non-symmetrical, never-ending variety, but presented here in test-tube-like ranks. This was also the period of the patronizingly infamous remark by the singer Taylor Swift, "I would do a ginger." (Try substituting "person whose skin, rather than hair, is a different color from mine" for "ginger" in that remark, and see how bad it tastes then.) But in 2014 the

Copenhagen Post reported that Cryos was now struggling to keep up with the demand for MC1R sperm. The turnaround was being driven by demand from couples both straight and gay in which one partner had red hair, so a redheaded child was desired to create the visual effect of consanguinity. And jolly good, too. But why, after centuries when society's response to red hair was mixed, to say the least, should it in our own day and age become desirable? Is it simply that appetite for rarity, as with those teenagers in Japan, where what was bad becomes good because it was bad? Or is red hair coming to stand for something other than the wimps, the barbarians, the kooks, the witches, the bluestockings, the sexpots?

There is this one question that our species has been asking itself ever since it stood upright: what are we? In our age, that too has undergone a subtle shift. We ask now: what am I? Our texts, our Facebook postings, our tweets to one another and to total strangers, people we have never met and never will, our blog entries, our websites, our Pinterests and Instagrams and uploads and likes and dislikes and all our multiple new means of communication are part of the attempt to answer that question. They're there, they're used, because we need them. In a way the network we create around ourselves duplicates our far-flung genetic connections to one another. And in all the words we now throw at one another, around and around the planet, day and night without ceasing, we are both all duplicating one another's behavior and all trying to define ourselves as apart, to give our likes and dislikes an individual value.

It isn't easy. We all see the same movies, are aware of the same celebrities and the details of their lives, have the same books, the same goods, the same fads and passing fancies thrown at us. Culture has become something that homogenizes us, rather than characterizes us, as it used to do. Against this background, perhaps red hair is starting to stand for something new and desirable. Perhaps it is starting to stand for individuality, for differentiation. And perhaps it is able to do this, even if twenty young actresses one after the other parade newly rubricated locks on the red carpet, because of the historic depth of its association with otherness and with the outsiders—the borderland, the liminal, the wildlings of society.

And perhaps the change in its status has something to do with redheads themselves.

After all, we know how to do this. If the decades since 1945 have taught us how to do anything, it is how you tackle prejudice head-on. You take hold of the stereotype used to define and to subjugate you, you confront and assess it. You reject its negative aspects. You go to war on them. The redhead's reputation for a fiery temper could come in very handy here. But you take the aspects of your image that you like, and you use them to fashion something positive and viable. For Conan O'Brien, one of the most successful, perspicacious, and quietly canny redheads of them all, that is his career. For many of us, it describes the arc of our life from childhood bullying to an adult sense of pride in our identity and our genetic inheritance.

Or at least it should. There have been some horrendous cases in recent years of redheaded children being bullied to the point of suicide. That has to stop. Would it be acceptable for a child to be bullied at all, let alone to death, because of their skin color? Their religion? Their own or their parents' race, their own or their parents' sexuality? Making it stop, making people more aware of the language they use and their attitudes toward red hair, was one of the impulses behind Thomas Knights's hugely successful *RED HOT 100* project. Redheads are also finding a voice for themselves. When in 2000 the UK energy company npower ran an ad featuring a redheaded "family" with the tag line "There are some things in life you can't choose," the flood of complaints against npower was dismissed. But things move on. The tag line was picked up and used by Eleanor Anderson, a redhead herself, as the title for her 2002 thesis on attitudes toward redheads. Two things to note here: that red hair should be the subject of a thesis, to begin with, and then the repurposing of this notorious ad to attack the very attitudes responsible for its creation. In Australia, where "ranga" is still supposedly an acceptable term for a redhead (hard to know who should be the most insulted by this, redheads or orangutangs), the Red And Nearly Ginger Association is doing the same thing. Picking up the pace, in 2014 the Australian company Buderim Ginger ran an "Australia's Hottest Ginger" competition as a marketing tie-in, and although some might say this was more a cynical reuse of a stereotype than a repurposing of it, it was not, refreshingly, for women only. There was a male hottest Australian ginger as well as a female. There are redhead websites, some of which, such as How to Be a Redhead, Everything for Redheads, and Ginger Parrot are models of how to e-market and thus by sleight of hand how to e-lobby, as well, as are Ginger with Attitude, Ingingerness, and Ginger Problems, all with a nice line in turning bias against red hair on its head. Bit by bit, attitudes start to waver, then to change. When in 2009 the Tesco supermarket chain in the UK offered a Christmas card showing a redheaded child sitting on Santa's knee with the legend "SANTA loves all kids. Even GINGER ones," the outcry and embarrassment to Tesco was such that the company not only apologized but withdrew the card from sale as well. So yes, attitudes are changing, but they never change fast enough. "Gingerism," so-called, is a truly ugly word for an ugly thing.

Part of the problem here is that gingerism doesn't *look* like it's racism, and in a way it's not, or at least not in the way we are used to thinking of it. Race may not be involved at all. And red hair does stand out, it can't help it, so for those of that befogged and bigoted understanding, in apparently calling attention to yourself it's as if you are asking to be picked on.[114] Then, redheads are not so other that we are going to turn out to be unexpectedly dangerous. We're a known quantity. Our skin is *white*. You see two people whose skins are different colors with one victimizing the other, you know what you're looking at. But with red hair there are likely to be none of those obvious clues between bully and victim; only the color of the hair. And who would pick on another human being just because of that?

But perhaps the most significant development is this: that redheads are no longer merely that 2 to 6 percent, those isolated individuals within the rest of society. Redheads are rapidly forming a community. There are redhead festivals in Russia, in Scotland, in Ireland, and one has even been started in Israel, on the aptly named Carrot Kibbutz. There are ginger pride events in the United States in Rome, Georgia; Portland, Oregon; in Chicago; in New York; and in Austin, Texas; in Milan, in Manchester in the UK, and in Montérégie,

---

114 See Druann Maria Heckert and Amy Best, *op. cit.*, for a more detailed discussion.

Quebec, in Canada. They are growing a new awareness of redhead identity and worth, of redheads' knowledge about themselves socially, scientifically, and culturally. There is naturally a sense of ganging together, of a rallying, of pride at these events. There is speculation of a redheaded moment not that far off, a redhead renaissance, indeed. And the biggest festival of them all is held in September in Breda, Holland.

REDHEAD DAYS

Hair ... mediates between the individual and their
culture. ... It is a site of immense conflict—external
authorities, parents, church, peer groups, school
gangs and fashion gurus all seek to impose
their conventions on the individual.
JULIET McMASTER, "TAKING CONTROL," 2002

What began as the colour of children, comics and
clowns is now a flag of determination.
GRANT McCRACKEN, *BIG HAIR*, 1995

The town of Breda stands where the River Aa broadens upon meeting the River Mark, about sixty-five miles south of Amsterdam. "Brede" means "wide," so the name of the town means "wide-Aa," basically, which is not only easy to remember but conjures up pleasing visions of an European Economic Community river-naming directive; one that might mean some other river, twenty-six water-courses distant, would be known only as the "Bb."

In its long history Breda has been bought, sold, taken in battle, inherited as a dowry, and (briefly, in 1795) French. It has been burned to the ground, immortalized by Velázquez, and laid siege to by the Spanish. In 1590 it was recaptured by the Dutch when a tiny force of sixty-eight men managed to get into the town by hiding under the turfs in a peat-boat, which is about as Dutch a version of a Trojan horse as you could ask for. Charles II lived in Breda during most of his exile and signed the Treaty of Breda in 1667, by which England gained the far-off territory of New York, but precious little else. Polish soldiers liberated the town in 1944. Such are the fortunes of war, however, that General Maczek, who led the liberating force, and who died at the age of 102 and is now buried in Breda, spent his old age working as a barman among the redheads of Edinburgh. The town is known for chocolate, lemonade, licorice, and beer. It has a park, the Valkenberg, and a splendid Grote Kerk, a building of such Gothic ribbiness, such piercings and knobbly crockets that it looks as if its bones are poking through its skin—a great gray Gothic pachyderm, quietly moldering away under those horizonless Dutch skies. But in all its long, eventful history, I doubt Breda ever anticipated any such happening as Redhead Days.

It is, after all, a very strange thing for a redhead to find him- or herself completely surrounded by other redheads. As the actress Julianne Moore has said, redheads notice one another, we become preternaturally alert to another redhead in the room—there's even a redhead look, a glance of complicity that passes between us. But put 6,000-odd redheads into the center of one small historic Dutch town, and complicity ain't in it. That's no longer a minority population of any sort, that's a tribe. Not bad for an event that started by accident, with the Dutch painter Bart Rouwenhorst placing an ad for models to act as inspiration. Rouwenhorst likes redheads. He paints redheads. In 2005 he wanted fifteen or so redheaded models for a sequence of paintings inspired by the works of Rossetti and Klimt; some 150 turned up. Faced with the choice between creating 150 paintings or a festival of redheads, Rouwenhorst went for the latter.

So here I am on Eurostar, heading for Brussels and ultimately to Breda and the largest gathering of redheads on the planet. I've been looking out for others of my rubified kin ever since manhandling my suitcase through security at St. Pancras, and thus far I haven't spotted a single one. I mean, I know we're rare, but. Six thousand redheads (the number the organizers of Redhead Days tell me they are expecting this year), at a very generous average of 6 percent of the population at large, would be the seasoning for 100,000 or so non-redheaded folks. Surely the number of people at St. Pancras should include one other, at least?

Apparently not. What my carriage can offer instead is a party of jolly sixty-somethings, on their way to Bruges. Their conversation is loud and joyous; it's one couple's anniversary, it's someone else's birthday, a fourth member of the group has just retired. They have crackers, pâté, wine. The men twit the women over the amount of chocolate being in Bruges will require them to eat. The women hoot with laughter, josh the men, and exchange sotto voce wisecracks of their own. The men are mostly dressed in affluent beige; the women are much more colorful—turquoise blues, purples, coral reds.

Red. I suddenly notice the queen of the group has hennaed hair. She has a voice so rich with damage (cigarettes, hard liquor, holding its own in God knows how many marital disagreements) that it almost carries its own static. The end of every sentence she utters, every tale she tells, is lost as the group dissolves with laughter, bending over their tabletops, lifting their plastic tumblers of wine up high as if to keep them safe. There's another woman seated beside her dressed in pale pinks and blues, round-faced, round-eyed, a little blonde duckling, tucked in beside this creaking, cackling scarlet parrot. *Red*, I find myself intoning inwardly, *is the color of dominance.*

The last time I was one among many redheads was as an extra, filming *Atonement*. A lot of titivating goes on in the longueurs between takes, and it passes the time, having your hat adjusted, lipstick renewed, 1940s pageboy recurled. "You have lovely hair," the woman with the curling tongs said, behind me, holding on to a hank of it with urethra-contracting tightness as all around us cables writhed across the floor like Laocoön's serpents, and the lighting guys cursed at the strength of the sun soaking into St. George's, Hanover Square. "Lovely color," she continued. "Is it natural?" She pulled the curling tongs free. "And really thick. Redheads always have lovely thick hair." I'm about to correct her—no, it's not thicker hair, it's thicker *hairs*, but seizing another hank, winding it around her tongs, she forestalled me. "It's a shame you're all going extinct. *Is* it your natural color?" The time before that was a family holiday near Balmoral, where everything was red—the deer, the squirrels, the grouse, the heads of hair on the people. We went to the Highland Games at nearby Ballater, and from a lifetime of being the odd one out in this family of blondies and brownies, suddenly I fitted in, and it was the other members of my family who were the anomalies, the Sassenachs, the ones who stood out from the crowd. Now here I am heading toward Redhead Days in Breda and there's not another redhead in sight.

I have to change at Brussels, wait for a train to Roosendaal.

I have to change again at Roosendaal for Breda. At Breda I am to speak, for an hour, to an audience of redheads about the history of the redhead, be interviewed by a documentary crew, and get myself photographed with a bottle of Gingerella ginger beer. These are not things I have ever done before, nor are they things I have ever envisaged myself doing. Even saving my receipts from this journey is a reminder that I am here in my capacity as a *professional* redhead. I am a tad nervous.

The men in beige, the scarlet parrot, and her blonde duckling girlfriend decant from the train still holding their plastic cups of wine on high. I locate the platform for Roosendaal, then go buy myself a beer.

And I'm standing there, on the concourse, drinking my beer and people-watching when I see a man, whitish hair in a long ponytail, denim shirt, South American striped and tagged and fringed and toggled waistcoat, fingers full of silver, and the reason I see him is because as he approaches, he is staring at me. He takes in my face, then his eyes go around my head as if checking on my aura, and as he passes, he pulls an imaginary hat from his head and tilts me a bow. "Enchanté."

Man with a Thing for Redheads.

On we go. I'm watching, on my iPhone, the blue dot that is me, traveling through these hinterlands of Belgium and on into the Netherlands, and through or past so many of the towns where the medieval artists, the medieval men with a thing for redheads lived. Brussels, where Rogier van der Weyden died. Bruges, which had Jan van Eyck. And now here we are at Antwerp, which had them all, including in the winter of 1885, Vincent van Gogh—cold, lonely, hungry, miserable, ill, ginger hair close-cropped as a prisoner's, sitting for hours in front of the paintings of Peter Paul Rubens (whose own surname in Latin means "colored" or "tinged with red"), worming his way into Rubens's paint swirls in his thoughts and bringing red into his own palette thereafter. The sky is boiling with clouds, the end of Hurricane Cristobal, flicking Europe with its tail. Strobes of sunlight pass across the fields, stride over the roof-scapes of the towns. Van Gogh was born near Breda. I feel as if I'm approaching redhead ground zero.

The train is much, much quieter than was the Eurostar. There's a discreet conversation in French bubbling like a water fountain to the right of me; I try to listen in as subtly as I can while taking notes. It seems one half of the conversation is recommending a hairdresser to her friend. I am reduced to the amateur subterfuge of examining both, as far as I can see them, in my handbag mirror. Writers are awful people, we respect nothing, and Sherlock Holmes would be proud of me. Sure enough, one purplish-reddish head, one orangey-brown. People have asked me all the way through the writing of this book, "Does dyed red hair count?" Of *course* it does. What more wholehearted acclamation could there be?

Roosendaal. Time to go grapple with the suitcase again.

Roosendaal is one of those stations where the platforms are as far apart as the coasts of a couple of continents, facing each other over a waste of rusty unused lines and wonky buddleia bushes, bright and lively in this Indian summer with bees. Traveling like this on my own to a place where I will be both new and everywhere, where I will know no one and everyone, is strangely nerve-inducing. My fellow redheads in Breda could include a little girl from the Kibbutz

Gezer (Carrot Kibbutz), a teenager from Afghanistan, and a Venus from Australia. It's ridiculous to think I am related to any of them in any way at all, or they to me, to presume that we will have anything in common, and yet we will. We do.

A train, grinding in. The sign hanging above the platform does a noisy electronic blink. BREDA.

There's something unshowy, self-effacing about Dutch towns, like the people—endlessly polite and, for a Brit, shamingly bilingual. They don't do look-at-me high-rise. Breda sits low to the landscape, the same height to the buildings as they have had for centuries. I do a little walkabout after checking in to my hotel (the Golden Tulip—how Dutch is that?) and register vague impressions of grayness and stone and mild hubbub from the bars. But the redheads, if they are here yet, are plainly all in hiding. I spotted two, in the park, from my taxi on the way to the hotel and that's it. Back at the hotel and the only thing orange is the Dutch football team on the television in the bar. I, like Breda, also have a Mark. I get into bed with my laptop and email mine assorted disconnected thoughts and a string of kisses.

According to my dog-eared program, I will be able this weekend to indulge in a weekend's worth of redheaded music, join a pub crawl, take a canoe trip or win one in a hot-air balloon, have my fortune told, my nails done, my colors done, get a makeover, speed-date, do a fire art workshop, a Lindy Hop workshop, a burlesque workshop, a cocktail workshop, buy more redheaded merchandise than

I ever knew existed, watch a catwalk fashion show, immortalize myself in the group photo, and learn how to keep my energy levels high by laughing at myself, which will no doubt come in very handy at two p.m. on Sunday afternoon. Before that, I have to meet with Papercut Films.

What the hell do you wear to be filmed? Solid colors, I am told, no black, no white, which leaves such an endless amount of space for picking the wrong thing you could fit my entire wardrobe into it. We start by walking me around the nave of the Grote Kerk, now bedecked with Thomas Knights's photographs, one of which apparently had to be removed when the church council took exception to the lowness of the model's low-slung jeans. The camera follows me, around and around the curve of the nave, and then behind me, up the winding stone stair to the room where the filming and interviewing proper is to take place. I manage to resist making the obvious and awful joke, as the cameraman labors up the stairs behind me, as to whether my bum looks big in this. I sit as directed and am powdered anew—apparently I'm shiny. The camera rolls. I keep such desperate eye contact with my interviewer that in the heat of the room my eyeballs start to dry out. Being interviewed seems to consist of being asked the same question three different ways until I give them an answer they can use, until Chris and Mark—another Mark?— are happy. I have a horrible sense that most of my answers are a disappointment to them, but I find an unexpected core of obstinacy within myself. Never mind the professional redhead, the redheaded activist has decided the time to assert herself is *now*. I have spent months surrounded by books and theses and journals and offprints, and in my opinion the true history of red hair is infinitely more fascinating than any of the myths. So no, redheaded young women were not hauled off by the hundreds to be burned at the stake as witches; no, we do not bleed more than other hair colors; no, we do not originate in Ireland; and no, we are not going to become extinct. The noise from the street outside is increasing. The crew is filming at the speed-dating session next, and there is some discreet checking of watches going on. The session wraps. There is lunch. I say that I hope I wasn't too useless. "You were fine," Chris tells me, very kindly. Music is bass-booming from a band in the town square. I head down the stairs, outside, and—

Whoa. *Whoa*. What the—what's the—whatever the collective noun is for six thousand redheads (a bonfire, a sunrise, a solar flare, a rubescence, an incarnadination, a conflagration, an incandescence, a frenzy, an *apocalypse* of redheads), let's hear it now. There are tall redheads, short redheads, plump redheads, thin; there are tiny little ones charging through the crowd at knee-height; there are redheads in baby carriers and redheads in wheelchairs. There are old red-heads, gone sandy with age; there are redheads so new and young that to see any hint of color on their infant heads would be an act of faith only two redheaded parents could perform. There's an open-top bus with redheads on its top deck, an immense poster of the crowd from last year's festival completely covering one side, and redheads having their photographs taken in front of it, a sort of past, present, and future of red. There are redheads in costume—Rapunzels, mermaids, Magdalenes, vampires, Vikings. There's the redhead queen from the Irish Redhead Convention here somewhere; there are redheads in facepaint: foxes and squirrels. A man passes me with a dog, a red setter with a spotted handkerchief tied around its neck, padding along at heel, which makes me laugh—do redheads chose red pets? Ginger cats? Pomeranians? Dachshunds?

And there is hair, hair, everywhere. There is hair here that has plainly never been cut once in its life. There are *manes* of hair. There are braids down to hips; there are walking bushels of red, there is every shade of it describable. Tight deep red curls; curtains of pale, pale cinnamon; orange-hued skinheads; terracotta plaited like lawn edging; peppery cornrows; masses of ginger so prodigious they follow their owners a bounce or so behind with beards and whiskers to match. There are mothers and redheaded daughters, bonding as only mothers and daughters will. There are tribes overlapping— rockers and bikers and punks circulating around the merchandise tents side by side with neat little families in Peter Pan collars with babies in strollers. There are redhead dreads. You stop looking at the people—the first thing you measure up on everybody is the hair, and there is a certain air of rivalry in some of those sizings-up as well, a slight sense of "Who's the reddest?" It certainly isn't me. There are redheads here who put me to shame. One of the questions Papercut asked was where do redheads feel they belong? If we can come from so many different places, where's home? Here's the answer—right here (Fig. 32).

I wander the streets, sketching tableaux. The one redheaded sibling, for example, in a group of three, where it's the other two (bigger, older) who look awkward, who are now the odd ones out. A redheaded couple, doing that linked-together couple's walk, arms across each other's backs and fingers through each other's belt loops. She has her head on his shoulder and from behind their hair is meshing, too. A teenage girl in a costume made of a cloud of purple netting (this year's signature color), through which her face peers as if through purple stage smoke. A giant red plait, draped from the upstairs window of a bar. An older couple, must be in their seventies, in matching sweatshirts and sturdy hiking boots, who look both totally at home here and totally out of place—her hair faded with time, clipped sensibly short, his almost gone. They'd have been toddlers when the Nazis were in town, which reminds me of the one internet factoid that has eluded all my attempts so far to run it to ground—that the Nazis forbade redheads to marry. I can't believe red hair was any different a signifier for the Nazis than it was for the bureaucrats of the Spanish Inquisition. The way prejudice works is never more than predictable.

And standing here in the square outside the Grote Kirk of Breda is surreal. It hadn't occurred to me before how much a part of being a redhead is the business of being the only one on the train. It's so much a given of having the hair that no one else has the hair, but here everyone does. Breda makes the outsiders the tribe; the exception the rule. So if we are no longer the other, who are we?

Ruth is here from Bristol with her sister and her two sons. I find her in the Grote Kirk, admiring the men in Thomas Knights's photographs. It's very easy to fall into conversation on Redhead Day in Breda. Everyone seems keen to share the experience of being here and of being red in a manner that has something to it of catharsis. Her sister is a redhead too, but a toned-down version; Ruth is fire-engine red. Her two boys are darker, she tells me, pointing to one of the photographs, "like him." Teenagers, off on their own; she suspects they will be down at the redhead speed-dating tent. Do they like redheads? She laughs. "They like girls." It's kind of telling, that the graphic for the speed-dating venue has a gorgeous redheaded coquette, fluttering her lashes at a man whose hair is brown. She frowns when I point this out. Does she think it's different for girls?

It turns out Ruth is an army wife; when her boys were born, the family was in Germany. Her sons went all the way though kindergarten and junior school without a problem; then they moved back to the UK and everything changed. "They were big enough to stand up for themselves then," Ruth says, and from the sound of it, just as well they were. Do they get treated differently now? "There's two of them. They look out for each other." It was the same with her and her sister when they were young. Harder to pick on two. And attitudes are changing; she's noticed that. Things are different now.

Are they?

A long, long pause for thought. "Yes," she says at last, and then brightens. "I mean, there's Prince Harry, isn't there?"

Marius's family is half Hungarian, half Romanian. Despite his youth (when I offer to buy him a beer, he modestly requests a Coke), he's a Redhead Days veteran. Heard about it through the internet in 2006—could any of this be happening without social media? Helped out as a volunteer last year, brought along a gang of friends this. He and his mates have also been down at the speed-dating tent, which must be doing a roaring trade. He's also been to one of the gay bars in town, which was a good experience, as he calls it. He is trilingual at least; has moved with his family to Scandinavia and then with them to Germany. What was it like being a redhead in the land of the Vikings, I ask, anticipating a positive response, but Marius pulls a face. He was bullied, he says, "a little bit" in Norway, but that was because he wasn't Norwegian. His remark puts me in mind of a depressing suggestion made in their study by Feinman and Gill, that people have an inherent psychological need to dislike *something*.[115] All the same (Marius is thoughtful when he speaks, weighing his words), on balance he thinks red hair for guys is a bad thing, despite the fact that if you have red hair it makes you more tolerant, more aware of the feelings of others. His role model is the character played by Tim Roth in *Lie to Me* (to whom, it must be said, he bears more than a passing resemblance). Where, having lived in so many countries, does he feel at home? "Here."

Just to confirm my worst suspicions of the prevalence of being bullied in so many redhead lives: Kelly. Kelly comes from New Zealand, is a media student, and has deep red hair that truly does shine like polished copper and a tale of being picked on and marginalized all the way through her school days, with one of her worst persecutors being the only other redhead at the school. "Only," says Kelly, "she wasn't as red as me." Now, at college, her red hair is both envied and emulated. This is indeed the story for most redheads: you're teased as a child (and even the most well-meaning of teases, from those you know are fond of you, is wearing and unwelcome, please note—those teased always taste the vinegar more strongly than the honey). Then you grow up, and especially if you're a girl, your expe-

---

115 Saul Feinman and George W. Gill, *op. cit.*

rience of being red can transform with bewildering rapidity. Are you happy now with being a redhead, I ask her, thinking what a tragedy, with hair as beautiful as that, if she's not, and she nods her head, vigorous emphasis: yes. Now, it's her. Everything about it is her—even the interest and awareness it has given her, as a media student, in what makes other people tick. Happy ending. So should they all be.

Laura-May Keohane, crowned queen of the Redhead Convention at Crosshaven in Ireland in August. She's here in a red-and-gold cloak, like an old-time coachman, gowned and crowned, one of the stars of the show, and is so ridiculously pretty that when people stop to take photographs of us I want to get out of the picture. There is a huge amount of snapping going on. Two redheads need only stand together to be photographed by someone. You get the feeling this is not simply building up an archive of memories; there's some dedicated search for self-definition going on here as well, a crowd-sourcing exercise in personal classification. Laura-May is astonishingly gracious in her role; if I had been crowned queen of the Irish redheads at the age of twenty-one I doubt I would have been quite so mature about it. "Now where does the red hair come from?" she asks, smiling away as the camera-phones are flourished all about us. She has a lovely line in Irish lyricism, too. "There was a fella told me it was caused by the clouds and the rain."

Dr. Tim Wentel, my fellow speaker in the Grote Kirk at Redhead Days. He's the man to tell you where red hair comes from, if anyone can, but as well as being an expert on the hair, the skin, the freckles, he's something of an authority on beers as well. My notes from our meeting stray exuberantly up and down the page and at one point toddle off into the gutter of my notebook for a lie-down. We end up at an Indian restaurant where I preserve the redhead's reputation for dealing coolly with the hottest curries, and Dr. Tim gives me chapter and verse on the origination of the redhead extinction story—predictably, another internet myth, from years ago, but as he tells it, with a twist. Globalization may, eventually, many generations down the line, succeed where the makers of woozles have failed. We may all, as we mix, eventually revert to the phenotype of dark hair, eyes, and skin with which we once emerged from Africa. Given global warming, of course, we may all be thankful that we do, but if globalization should have this effect, *if*—does that mean no more red hair? Anywhere? Or might Mother Nature have more tricks up her sleeve?

I'm up at the business end of things now, tents selling t-shirts and wristbands, makeovers and do-overs, and it is beginning to feel a little odd, this relentless concentration on just the one body part. I have an urge to stand on the steps to Breda's noble town hall and shout, "I am not a redhead! I am a free woman!" I tell myself I at least will no longer gawp and stare at my fellow redheads. Instead, I step away and almost immediately find myself on the outskirts of a small circle of people gawping in amazement at a tall teenage girl who not only has the most wonderful skin, like illuminated bronze, but with it the most astonishing red, red, red, *red* afro, an aureole as bright as molten lava. This, so I learn, is Sterra. And Sterra is what Mother Nature can come up with if you only give her the right breaks (Fig. 33).

Sterra is here at Breda with her mother, Irmgard. They've been coming here since Sterra was a child. And while Irmgard's hair is now dark, when she was young, she tells me, it was red. Irmgard is Dutch, but Sterra's father is from Senegal, and somehow those two gene pools, as they mixed, created this thirteen-year-old standing here. I ask the obvious question: has Sterra ever experienced any racial prejudice (wondering, even as I ask the question, what on earth, given her coloring, the form of such prejudice could take)? No, Irmgard assures me, never. There's a line in Emily Cameron Walker's thesis to the effect that red hair is no more than a "genetic spandrel," that pale skin is the genetic driver, red hair the side effect, and technically yes, she is right. But in that case, looking at Sterra, all I can say is *quelle spandrel.*

Sunday. My talk is mere hours away. I'm brunching in Breda's main square, fortifying myself, and David and Anna are at the table next to mine. Anna has princess ringlets of the palest ginger, delicate as a cobweb, all the way down to her waist. She also has a wonderful story from her days at school. The rest of her class developed a game they called "carrot-busters" where every break time they would pile on top of or cannonade into Anna and the two or three other redheads in the playground. (Months later I'll find myself talking with a class of embryonic writers in a school in New York. The members of the class are about the age Anna was in her story, and we start talking about discrimination and prejudice and red hair. They all mention one girl in their year who has red hair, and thinking of Anna and her carrot-busters game, I ask if their classmate has ever been picked on in any way, and the whole of my audience recoils in horror. And that, I guess, is the difference, between growing up on the Upper West Side and growing up in the West Country.)

I turn to David and ask him if he noticed Anna because of her hair, if he would call himself an example of Man with a Thing for

Redheads. "Well," he says proudly, "Anna is the only girl I've ever asked to marry me." And Anna waggles at me on her left hand the most delicate diamond ring, so unmistakably new that even the hand has the look of still getting used to it.

The Grote Kirk in Breda can seat one thousand people, and I'm not saying it's full, I'm just saying that from the podium set up for us speakers it certainly looks that way. Row upon row of people fanning themselves with postcards of Alexa Wilding. I've been warned by the organizers that given the sensibilities of those who care for the Grote Kerk, I should mention that there will be nudity in the slides I'm going to be showing. So I'm now scrawling WARN THEM ABOUT THE COURBET in caps two inches high across the top of the first page of my talk when the hum and scrape of people settling into their seats is drowned out by a motorbike roar from outside. I look up. For the first time it registers quite how much studded leather as well as hair there is among my audience. Well, well. It seems I will shortly be explaining the historical significance of *L'Origine du Monde* to a gang of bikers. This is going to be interesting.

Joachim, who has gallantly volunteered to video-record my talk, gives me the thumbs-up. I take a deep breath.

"Okay. Ladies and gentlemen. Let me introduce you to King Rhesos of Thrace."

So after all that, where are we now in the history of the redhead?

First, I think Ruth is right. I think attitudes are changing, and I think it is redheads who are changing them. When, in response to the notorious *South Park* episode, "Kick a Ginger Day" was set up on Facebook by a particularly misguided fourteen-year-old in Vancouver, the Canadian comedian and activist (and redhead) Derek Forgie boldly co-opted the Canada Dry logo and pushed back by setting up, also on Facebook, "Kiss a Ginger Day." Nothing like turning your opponents' weapons against them. This year Kiss a Ginger Day became such a phenomenon it was trending on Twitter, although it certainly didn't hurt that it fell just before the announcement of the nominations for the 2015 Oscars. Lo and behold, redheads were so in the camera's eye that in the UK the *Mirror* newspaper ran the ebullient headline "It's Kiss a Ginger Day! Here are 13 red-haired celebs we definitely want to celebrate with," and went on to list them: Damian Lewis, Michael Fassbender, Emma Stone, Karen Gillan, Benedict Cumberbatch, Christina Hendricks, Prince Harry, Amy Adams, Eddie Redmayne, Ed Sheeran, Isla Fisher, Rupert Grint, and Lily Cole. Here is how you change a stereotype: you make it cool. You take what was marginalized and you make it desirable just by pointing out how unusual it is. You turn its downside upside. Simple. Comedians such as Catherine Tate get in on the act, with her "Ginger Refuge" sketch (270,000 views on YouTube and counting). Thomas Knights launches his *RED HOT 100* photographic show, to "rebrand the ginger male," as its creator puts it, and takes it around the world. Papercut Films creates their documentary, the catalyst being the experience of my interviewer, growing up as a boy with red hair. And in Breda, after his talk, I watched couple after couple come up to Tim Wentel, wanting a forecast of their chances of having a child with red hair, and asking this as something which they, like the clients of Cryos International, positively desired.

And this, I suspect, is the real Redhead Dilemma, and it doesn't have a thing to do with invisible eyebrows. We want to shake off the perjorative associations of being red, but we don't want to give up our so-called rare color advantage, the thing that makes us stand out, that sees us exchange the redhead look in public, that means we feel special, rare, unique. We want to have our ginger cake and eat it. We love Ed Sheeran, and that he calls his red hair his saving grace; we adore the fact the Michael Fassbender grows his ginger sideburns long; at the same time we wish to God that Putin weren't so carroty. We groaned in disbelief at *I Wanna Marry Harry* when that appeared on our television sets (to be fair, so did everyone else), yet when Ruth Wilson, Amy Adams, and Julianne Moore all triumphed at the Golden Globes and Julianne Moore went on the win the first redhead Best Actress Oscar, it felt like a tipping point. And I don't think this is something where you can pick and choose. If we were ever in a world where redheads *weren't* singled out for the color of their hair, where that wasn't the one thing about us that everyone remembers, would we really like it more? Looking at so many redheads, all with such different stories, let alone with such different reds, brings it home all the more forcefully. Redhead Days is a celebration of individuality just as much as it is of our one uniting factor. To see every redhead as being the same as every other is absurd.

And that goes for us all. I began working through the final draft of this book in New York, watching thousands of people march wearing t-shirts bearing the words *I Can't Breathe*. I ended it watching more silent marchers, this time wearing t-shirts reading *Je Suis Charlie*. And if all this seems to have become rather political all of a sudden, that's because when you drill down into it, dammit, it is. Sometimes it feels as if what will finish us off as a species is not climate change, is not running out of fossil fuels, is not some super-plague, is not even our deleterious habit of trashing the planet we live on. What will do for us in the end will be just two things: ignorance and intolerance. A world that can't deal with something as small and insignificant as people whose hair is a different color is one where there is little hope of dealing with any of the problems created by those far bigger issues, of different skins, different faiths, different loves, different lives. It's not simple at all.

But who wants to live in a world where we don't try?

READING FOR REDHEADS

Some of the works I have used in researching and writing this book really could have been cited on almost every page. I hope I have credited their insights wherever I have made use of them. For those readers interested in pursuing their own researches, these are the works, along with those footnoted in the text, that were of most value to me.

Eleanor Anderson: "There Are Some Things in Life You Can't Choose . . . : An Investigation into Discrimination Against People with Red Hair," *Sociology Working Papers*, (2002).

Ruth Barton (ed.): *Screening Irish-America: Representing Irish-America in Film and Television* (Dublin, Irish Academic Press, 2009).

Beth Cohen (ed.): *Not the Classical Ideal-Athens and the Construction of Other in Greek Art* (Leiden: Brill's Scholars' List, 2000).

Michelle A. Erhardt and Amy M. Morris (eds.): *Mary Magdalen, Iconographic Studies from the Middle Ages to the Baroque* (Brill, 2012).

Susan Haskins: *Mary Magdalen: Myth and Metaphor*, (New York: Riverhead, 1995).

Druann Maria Heckert and Amy Best: "Ugly Duckling to Swan: Labeling Theory and the Stigmatization of Red Hair," *Symbolic Interaction* 20, no. 4 (1997): 365–84.

Noel Ignatiev: *How the Irish Became White* (New York: Routledge, 2008).

Benjamin Isaac: *The Invention of Racism in Classical Antiquity* (Princeton, NJ: Princeton University Press, 2006).

Sandra R. Joshel: *Slavery in the Roman World* (Cambridge: Cambridge University Press, 2010).

Tara MacDonald: "Red-headed Animal: Race, Sexuality and Dickens's Uriah Heep," *Critical Survey* 17, no. 2 (2005): 48–62.

Catherine Maxwell: *Swinburne* (Plymouth: Northcote House, 2004).

Grant McCracken: *Big Hair: A Journey into the Transformation of Self* (New York: Overlook Press, 1996).

Juliet McMaster: "Taking Control: Hair Red, Black, Gold, and Nut-Brown" in *Making Avonlea*, ed. Irene Gammel (Toronto: University of Toronto Press, 2002).

Ruth Mellinkoff: *Outcasts: Signs of Otherness in Northern European Art of the Late Middle Ages* (Berkeley, Los Angeles, and Oxford: University of California Press, 1993).

Marion Roach: *The Roots of Desire: The Myth, Meaning, and Sexual Power of Red Hair* (New York: Bloomsbury, 2005).

Anthony Synnott: "Shame and Glory: A Sociology of Hair," *The British Journal of Sociology* 38, no. 3 (September 1987): 381–413.

Kelly L. Wrenhaven: *Reconstructing the Slave: The Image of the Slave in Ancient Greece* (London: Bristol Classical Press, 2012).

Kelly Wrenhaven, "A Comedy of Errors: The Comic Slave in Greek Art," in *Slaves and Slavery in Greek Comic Drama*, eds. Ben Akrigg and Rob Tordoff (Cambridge: Cambridge University Press, 2013).

ACKNOWLEDGMENTS

Without some kind of organization, this list of names could stretch out to an extent that would be simply embarrassing, so . . .

At Black Dog & Leventhal, the wonderful J.P., and the just as wonderful Becky, Pam, Kara, Maureen, and Stephanie; also Becky Maines ("Red Becky") and Andrea Santoro, Rena Kornbluh, Mike Olivo, Christopher Lin, Ankur Ghosh, Nicole Caputo, Cindy Joy, and (for his splendid redhead map) Stefan Chabluk. At Fox and Howard: Chelsey and Charlotte. Jonathan Clements and Barbara Schwepcke—in at the birth.

The staff of the British Library and of the LSE Library; of the National Portrait Gallery and of the Royal Collection Trust; indeed all the friends and colleagues, past and present, who have been kind enough to interest themselves in this. It would have been a lesser book without you.

For advice and assistance: Nikolay Genov; Jeroen Hindriks; J.T. Leedson; Yvette Leur; Chris, Mark and Sara of Papercut Films; Professor Jonathan Rees; Joe Schick; Karin Schnell; Kirsty Stonnell Walker; Julia Valeva; Irmgard and Sterra Vlamings; and Dr. Tim Wentel. Like poor Ralph Holinshed, battling the exigencies of deadlines, I can only say that I have done what I could, not what I would, and any errors are mine. Heartfelt thanks also to Thomas Knights, to Bart and all the staff of Redhead Days in Breda, and all those many redheads who responded to the idea of this book with such enthusiasm and have offered details of their lives and experiences to me with such generosity.

For Millie, for listening so patiently. For being there: my family, especially Alice, Sam, Emma, Jack, and Ellie, and of course Nick.

And for all the support, the encouragement, the patience, and the wisdom any writer could ask for—Mark. Neither this, nor its writer, would be here without you.

ART AND PHOTOGRAPHY CREDITS

Frontispiece: Stefan Chabluk

Fig. 1: Copyright © REZA/Webistan

Fig. 2: Gusjer, via Flickr

Fig. 3: http://www.healthcare2point0.com

Fig. 4: Copyright © Nikolay Ivanov Genov

Fig. 5: Digital image courtesy of the Getty's Open Content Program

Fig. 6: © The Trustees of the British Museum

Fig. 7: © Bayerisches Nationalmuseum München. This object is a permanent loan from Bayerische Staatsgemäldesammlungen.

Fig. 8: *Calvary*, 1475 (oil on panel), Messina, Antonello da (1430–79)/Koninklijk Museum voor Schone Kunsten, Antwerp, Belgium/© Lukas—Art in Flanders VZW/Photo: Hugo Maertens/Bridgeman Images

Fig. 9: *The Coronation of the Virgin*, completed 1454 (oil on panel), Quarton, Enguerrand (*c*. 1410–66)/ Musee Pierre de Luxembourg, Villeneuve-les-Avignon, France/Bridgeman Images

Fig. 10: *The Last Judgement*, 1473 (oil on panel), Memling, Hans (*c*. 1433–94)/© Muzeum Narodowe, Gdansk, Poland/Bridgeman Images

Fig. 11: *Altarpiece of the Dominicans: Noli Me Tangere, c.* 1470–80 (oil on panel), Schongauer, Martin (*c*. 1440–91) (school of)/Musee d'Unterlinden, Colmar, France/Bridgeman Images

Fig. 12: *The Crucifixion* by Jan van Eyck (1390–1441)/De Agostini Picture Library/Bridgeman Images

Fig. 13: Copyright reserved

Fig. 14: Courtesy Southern Methodist University

Fig. 15: Courtesy of the British Library

Fig. 16: Queen Elizabeth I, *c.* 1575 (oil on panel), Netherlandish School (16th century)/National Portrait Gallery, London, UK/De Agostini Picture Library/Bridgeman Images

Fig. 17: Queen Elizabeth I (1533–1603) being carried in Procession (*Eliza Triumphans*) *c.* 1601 (oil on canvas), Peake, Robert (fl. 1580–1626) (attr. to)/Private Collection/Bridgeman Images

Fig. 18: *Symphony in White, No. 1: The White Girl*, 1862 (oil on canvas), Whistler, James Abbott McNeill (1834–1903)/National Gallery of Art, Washington DC, USA/Bridgeman Images

Fig. 19: *Le Sommeil*, 1866 (oil on canvas), Courbet, Gustave (1819–77)/Musee de la Ville de Paris, Musee du Petit-Palais, France/Bridgeman Images

Fig. 20: *Portrait of Algernon Charles Swinburne* (1837–1909) 1867 (oil on canvas), Watts, George Frederick (1817-1904)/National Portrait Gallery, London, UK/Bridgeman Images

Fig. 21: *Beata Beatrix* (oil on canvas), Rossetti, Dante Gabriel Charles (1828–82)/Birmingham Museums and Art Gallery/Bridgeman Images

Fig. 22: *Found, c.* 1869 (oil on canvas), Rossetti, Dante Gabriel Charles (1828–82) / Delaware Art Museum, Wilmington, USA/Samuel and Mary R. Bancroft Memorial/Bridgeman Images

Fig. 23: *Ellen Terry as Lady Macbeth* by John Singer Sargent/Tate Britain

Fig. 24: *The Knight Errant* by Sir John Everett Millais/Tate Britain and *The Martyr of the Solway*, 1871 (oil on canvas), Millais, Sir John Everett (1829–96)/© Walker Art Gallery, National Museums Liverpool/Bridgeman Images

Fig. 25: *Combing the Hair (La Coiffure), c.* 1896 (oil on canvas), Degas, Edgar (1834–1917)/National Gallery, London, UK/De Agostini Picture Library/Bridgeman Images

Fig. 26: *Cora Pearl* ©National Portrait Gallery, London

Fig. 27: *Uriah Heep*/Wikimedia Commons

Fig. 28: Tintin Press Club

Fig. 29: © Danita Delimont/Alamy

Fig. 30: Photograph by Pamela Tartaglio, courtesy of the Hollywood Museum in the Historic Max Factor Building. (PamelaTartaglio.com)

Fig. 31: *The Beethoven Frieze: The Longing for Happiness*, 1902 (mural), Klimt, Gustav (1862–1918)/Osterreichische Galerie Belvedere, Vienna, Austria/De Agostini Picture Library/E. Lessing/Bridgeman Images

Fig. 32: Picture by Colinda Boeren at the Redhead Days Festival in the Netherlands, www.redheaddays.nl.

Fig. 33: Photo made by Yvette Leur

INDEX

Also Available by Jacky Coliss Harvey

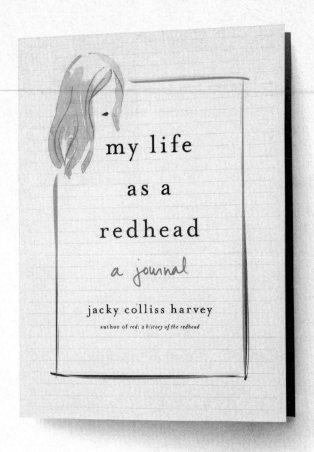

my life

as a

redhead

a journal

jacky colliss harvey

author of *red: a history of the redhead*

Exclusively for redheads!

A guided journal to record thoughts, feelings,
special events, and important decisions particular to the redheaded
experience. Every page offers a new, creative,
and thought-provoking exercise or activity to complete.